C. E. Fessenden

Elements of Physics

C. E. Fessenden

Elements of Physics

ISBN/EAN: 9783337277444

Printed in Europe, USA, Canada, Australia, Japan

Cover: Foto ©berggeist007 / pixelio.de

More available books at **www.hansebooks.com**

ELEMENTS OF PHYSICS

BY

C. E. FESSENDEN

PRINCIPAL OF THE COLLEGIATE INSTITUTE, PETERBORO', ONTARIO

London
MACMILLAN AND CO.
AND NEW YORK
1892

All rights reserved

CONTENTS

CHAPTER I
PAGE
MATTER AND ITS PROPERTIES 1

CHAPTER II
KINEMATICS 53

CHAPTER III
DYNAMICS 64

CHAPTER IV
HEAT 179

ELEMENTS OF PHYSICS

CHAPTER I

MATTER AND ITS PROPERTIES

INTRODUCTION

Means by which we obtain a Knowledge of Nature.—We are provided with means of gaining a knowledge of the things about us. We hear sounds, we see light, we smell odours, we feel heat, we feel the contact of other things, and we taste. Thus we have six gates through which information concerning the outside world may reach us. These gates are called *senses*, and any information received through one of them is called a *sensation*. Moreover, we have the power of reasoning; and through our senses, aided by reasoning, we acquire what knowledge of nature we can.

OUR SENSES SOMETIMES DECEIVE US

It must not be forgotten that we are often deceived by our senses, or, it may be, by our reasoning. We believe we have seen what we have not seen, or have heard what we have not heard. Thus to nearly every one, perhaps to every one, the full moon appears larger when near the horizon than when nearly overhead. You may prove that this is an illusion by holding a threepenny piece at arm's-length, nearly in a line between your eye and the moon in each of the two positions. Again, when that beautiful meteorite shot across the Canadian sky in June 1884, many who saw it affirmed afterwards that they had heard it whiz, while others, having quite as acute sense of hearing, experienced no such sensation. Doubtless the former were deceived through mistaking the meteorite for a rocket. As they had always heard a rocket whiz, they *inferred* that the meteorite did likewise. It is proverbial that a story changes wonderfully as it passes from mouth to mouth, and this has been set down as the effect of moral depravity in mankind. But it is rather the effect of mixing up *inferences* with *sensations*. The student of nature, if he is to succeed, must learn, among other things, not to allow unconscious inference to take the place of observation.

ORDER OF NATURE

The most *careless* observer of nature must notice that some events take place in a regular order, and that some causes are always followed by the same effect; he must see, in short, that there is an *order of nature*, and that all things do not happen by chance. The *careful* observer of nature, aided by the labours of those who have gone before him, cannot but believe that *nothing happens by chance*. It is the province of science to investigate this order of nature. Much has been done, but much more remains to be done.

A LAW OF NATURE

When we have discovered a fact concerning the order of nature, we call a simple statement of this fact a *law of nature*. We state a law of nature as fully and precisely as we are acquainted with the facts of which it is a statement. Further investigation may lead us to amend it, or even to substitute another for it. Thus, long ago, "Nature abhors a vacuum" was given as a law of nature, and was based on such observations as had been made up to that time. But further observations, of which you will learn more hereafter, led men to modify it.

A LAW OF NATURE IS NOT A CAUSE

It is necessary to bear in mind that *a law of nature is not a cause*. When Newton, from seeing an apple fall in his orchard, was led, from further observation and powerful reasoning, to enunciate the sublime law of gravitation, he did not discover the *cause* of the falling. No one since Newton's time has discovered the cause of a tendency on the part of bodies to move toward one another, and it may be doubted whether any one will ever discover that cause.

MATTER AND PHENOMENA

It has been said that any information of the world about us that we receive through our senses is called a sensation. That which, acting on our sensiferous organs, may produce a sensation, is called a *natural phenomenon*. Again, the source of all natural phenomena is *matter*. You have a sensation that causes you to say that you see a red-hot ball. The light which, acting on the sensiferous organs, produces the sensation, is a natural phenomenon. The ball itself, which is the source of the light, is a portion of matter.

MATTER MUST BE DISTINGUISHED FROM PHENOMENA

You must be careful not to confound matter with any phenomenon of which it is the source. Sound is not matter; but a bell, or anything else that may be the source of sound, is matter. Light is not matter; but the sun, or anything else that may be the source of light, is matter. An odour is not matter; but a piece of musk, or anything else that may be the source of odour, is matter. A flavour is not matter; but a piece of cinnamon, or anything else that may be the source of a flavour, is matter. Heat is not matter; but the flame of a candle, or anything else that may be the source of heat, is matter. Pain is not matter; but a sharp thorn, or anything else that may be the source of pain, is matter.

§ 1. **Experimentation.**—We observe phenomena which arise from natural conditions—that is, from conditions with which man has nothing to do, and we may learn much from such observations if they are carefully attended to. But we may have some doubt concerning the exact conditions under which a certain phenomenon arises. We then bring about accurately known artificial conditions, and observe the phenomenon arising. In this way our knowledge becomes more *accurate* or *scientific*.

It is a matter of common observation that water sometimes freezes, and that it freezes when the temperature is low. By trial we may ascertain the exact conditions under which water changes into ice. Whenever we place natural objects under certain accurately known artificial conditions, for the purpose of observing the resulting phenomenon, *we make an experiment*. It is, of course, necessary that we know exactly the conditions present in any experiment, or our trial, instead of teaching, will deceive us. It is not an easy task to make an experiment, and properly to read the lesson that it teaches, because it is very difficult to exclude all conditions whose presence we do not desire; and often conditions are present without our knowledge. When we wish to study a particular agent or cause we should so arrange our experiments as to lead to results depending upon this cause alone; or, if this is impossible, so that the effects of this cause shall so far exceed the effects of other causes that to neglect the latter will not lead to any serious error.

An experiment is a question put to nature. We receive the answer by means of a *phenomenon*—that is, a change which we observe, sometimes by the sight or hearing, sometimes by other senses. In every experiment certain facts or conditions are always known; and the inquiry consists in ascertaining the facts or conditions that follow as a consequence. The following experiments and discussions will illustrate :—

§ 2. **Things known and Things to be ascertained.**—We are certain that we cannot make our right hand occupy the same space with our left hand at the same time. All experience teaches us that *no two portions of matter can occupy the same space at the same time.* This property which matter possesses of excluding other matter from its own space, is called *impenetrability*. It is peculiar to matter; nothing else possesses it. These facts being known, let us proceed to put certain interrogatories to nature. Is air matter? Is a vessel full of air a vessel full of nothing? Is it "empty"? *Can matter exist in an invisible state?*

Fig. 1.

Experiment 1.—Float a cork on a surface of water, cover it with a tumbler, or tall glass jar, and thrust the glass vessel, mouth downward, into the water. In case a tall jar (Fig. 1) is used, the experiment may be made more attractive by placing on the cork a lighted candle. State *how* the experiment answers each of the above questions, and what evidence it furnishes that air is matter; or, at least, that air is like matter.

Experiment 2.—Hold a test-tube for a minute over the mouth of a bottle containing ammonia-water.

Hold another tube over a bottle containing hydrochloric acid. The tubes become filled with gases that rise from the bottles, yet nothing can be seen in either tube. Place the mouth of the first tube over the mouth of the second, and invert. Do you see any evidence of the presence of matter? Was this matter in the tubes before they were brought together? If not, from what was it formed? Which one of the proposed questions does this experiment answer? How does the experiment answer it?

Again, we are quite familiar with the fact that matter exerts a downward pressure on things upon which it rests; and that matter, in a liquid state at least, exerts pressure in other directions than downward, as, for instance, against the sides of the containing vessel. *Does air exert pressure?*

Fig. 2.

Experiment 3.—Thrust a tumbler, mouth downward, into water, and slowly invert. You see bubbles escape from the mouth What is this that displaces the water, and forms the bubbles? When the tumbler becomes filled with water, once more invert, keeping its mouth under the surface of the water, and raise it nearly out of the water, as in Fig. 2. Why does the water not fall out? What would happen if you were to make a hole in the bottom of the tumbler? Make the experiment with a glass funnel, first closing the small end with the finger, and then removing it. What conclusions do you draw from these observations?

Experiment 4.—Pass a glass tube through the stopper of a bottle (Fig. 3). Attach a rubber tube to the glass tube. Exhaust the air by "suction" from the bottle; pinch the rubber tube in the middle, insert the open end into a basin of water, and then release the tube. What takes place? Why does not the water fill the bottle?

Finally, we know that matter has weight, and nothing else has it. *Has air weight?*

Experiment 5.—Exhaust the air by means of an air-pump from a hollow globe (Fig. 4). Having turned the stop-cock to prevent the entrance of air, carefully balance the globe on a scale-beam, as in Fig. 5. Afterwards turn the stopcock and admit the air. What is the result? What does it teach?

The experiments with *air* teach us that *it is matter*, *since, like matter, it can exclude other matter from the space it occupies, it exerts pressure, and has weight*, while all the above experiments draw from nature one reply, MATTER CAN EXIST IN AN INVISIBLE STATE.

§ 3. **Minuteness of Particles of Matter.**—Physiologists teach us that, in order to smell any substance, we must take into our nostrils, as we breathe, small particles of that substance which are floating in the air. The air, for several metres around, is sometimes filled with fragrance from a rose. You cannot see anything

Fig. 3.

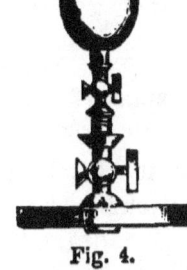

Fig. 4.

Fig. 5.

in the air, but it is, nevertheless, filled with a very fine dust that floats away from the rose. The odour of rosemary at sea renders the shores of Spain distinguishable long before they are in sight. A grain of musk will scent a room for many years, by constantly sending forth into the air a dust of musk. Though the number of particles that escape must be countless, yet they are so small that the original grain does not lose perceptibly in weight.

The microscope enables us to see, in a single drop of

stagnant water, a world of living creatures, swimming with as much liberty as whales in a sea. The larger prey upon the smaller, and the smaller find their food in the still smaller, and so on, till the power of the microscope fails us. The whale and the minnow do not differ more in size than do some of these animalcules, the largest of which are hardly visible to the naked eye. But as the smallest of these perform very complex operations in collecting and assimilating food, we must conclude that they are composed not only of many particles, but of many kinds of matter. These minute living forms that people the microscopic world are exceedingly large, in comparison with the inconceivably minute particles into which it is evident that matter can be subdivided.

§ 4. **The Molecule.—Experiment 1.**—Examine carefully a drop of water with the naked eye, or with a microscope. So far as you are able to see, the space occupied by the drop is entirely filled with water. Fill a test-tube with water (Fig. 6). Insert a cork stopper, pierced with a glass tube ; heat over a lamp-flame, and note the phenomena produced. Describe the result. Place it in ice-water. What happens ? Repeat this experiment, using other liquids, and compare the results.

This change of volume can be explained only on one of two suppositions : the space occupied by the water may, as it appears, be full of water, which the heat causes to expand, and occupy a greater space, as represented graphically in Fig. 7 ; or the body of water may consist of a definite number of distinct particles, which we will call *molecules* (as represented in Fig. 8), separated from one another by spaces so small as not to be perceptible, even with the aid of a microscope.

Fig. 6.

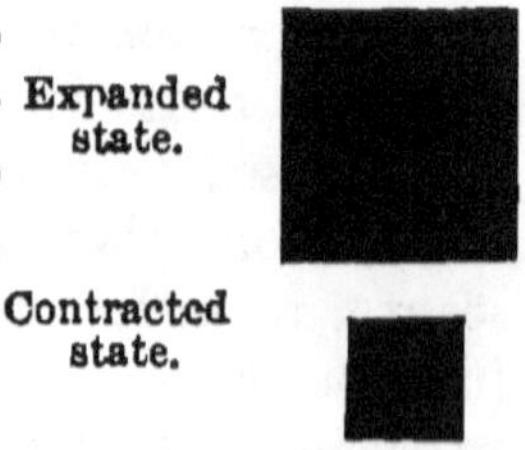

Expanded state.

Contracted state.

Fig. 7.

Expansion, in this case, is accounted for by a simple separation of molecules to greater distances. *There is no increase in the number of molecules, no increase in their size, only an enlargement of space between them.* Which of these suppositions is the more probable?

Experiment 2.—Place a tumbler full of cold water in a warm place, and in about an hour examine it. You find many small bubbles of air clinging to all parts of the interior surface of the glass. Is it probable that outside air has descended into the liquid?

Expanded state.

Experiment 3.—Place a tumbler half full of water under a glass receiver of an air-pump (p. 158), and exhaust the air. When a very good vacuum has been obtained, bubbles of air will be seen to form at all points in the liquid, and to rise and burst near the surface.

Contracted state.

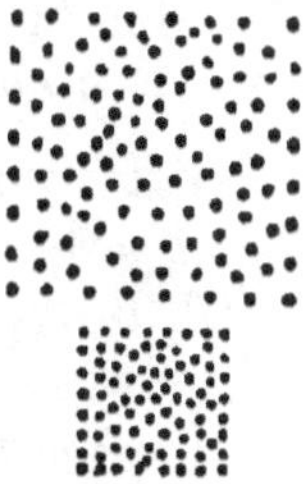

Fig. 8.

Evidently the air was previously in the same space occupied by the water. This seems to contradict the first of the above suppositions; for, according to that, the space occupied by the water is *full* of water, leaving no room for other matter. But according to the second supposition, the space is not *filled* with water; there is still room for particles of other matter in the spaces among the molecules of water. Now, as we cannot conceive of two portions of matter occupying the same space at the same time (*e.g.* where air is, water cannot be), we conclude that the glass "full of water" is not full of water. In a similar manner, it may be indicated that no visible body completely fills the space enclosed by its surface, but that there are spaces in every body that may receive foreign matter. If there are spaces, then the bodies of matter that our eyes are permitted to see are not continuous, as space is continuous, but every visible body is an aggregation of a countless number of separate bodies.

Perform at your homes the two following experiments:—

Experiment 4.—Pulverise one-half of a teaspoonful of starch, and boil it in two tablespoonfuls of water, stirring it meantime. What phenomena occur? What do they teach? What becomes of the water?

Experiment 5.—Fill a bowl half full with peas or beans. Just cover them with tepid water, and set away for the night. Examine in the morning. What phenomena do you observe? Explain each.

Strictly speaking, are bodies of matter impenetrable? What only is impenetrable? When you drive a nail into wood, do you make the two bodies occupy the same space at the same time? Do the wood and the iron occupy the same space? How only can you explain this phenomenon, consistently with the principles of impenetrability of matter?

§ 5. Theory of the Constitution of Matter.—For reasons which appear above, together with many others that will appear as our knowledge of matter is extended, physicists have generally adopted the following theory of the constitution of matter. *Every visible body of matter is composed of exceedingly small particles, called molecules; in other words, every body is the sum of its molecules. No two molecules of matter in the universe are in permanent contact with each other. Every molecule of a body is separated from its neighbours, on all sides, by inconceivably small spaces. Every molecule is in quivering motion in its little space, moving backward and forward among its neighbours, and rebounding from them. From some cause, which has not been explained, the molecules of a body have a tendency to rush together. When we heat a body we simply cause the molecules to move more rapidly through their spaces; so they strike harder blows on their neighbours, and usually push them away a very little; hence, the size of the body increases.*

This theory seems, at first, little more than an extravagant guess. But if it shall be found that this theory, and no other theory that has been proposed, will enable us to account for most of the known phenomena of matter, then we shall be content to adopt it till a better can be produced.

It may be well here to draw attention to the great difference between a *law of nature* and a *theory*.

A LAW OF NATURE is a statement of a truth regarding the order of nature, which has been learned from observed facts. But it is one thing to know a general fact, and quite another to know the cause of that fact.

HYPOTHESIS AND THEORY.—When a law has been discovered by careful study of observed facts, we next *imagine a cause*. We try to imagine a condition of things such that if it existed the observed facts would necessarily follow. If we succeed in imagining such a condition of things, we suggest an *hypothesis*. We next test this hypothesis in every way we can think of, and, if we can find no facts with which it is inconsistent, we then call it a *theory*. The more varied and apparently disconnected the phenomena which an hypothesis will reasonably explain, the more likely is it that the hypothesis is, in the main, correct. The student of nature should be careful to avoid the hasty enunciation of any hypothesis. No attempt should be made in this direction without as complete a knowledge of the facts as it is possible to acquire. The successful framers of hypotheses have been indeed few.

It is a very valuable exercise frequently to require the student to distinguish those statements in the text-book that depend upon theory from those that depend *only* upon observed facts.

§ 6. **Porosity.**—If the molecules of a body are nowhere in absolute contact, it follows that there are unoccupied spaces among them which may be occupied by molecules of other substances. These spaces are called *pores*. Water disappears in cloth and beans. It is said to penetrate them; but it really enters the vacant spaces or pores between the molecules of these substances. All matter is porous; thus water may be forced through solid cast-iron, and dense gold will absorb the liquid mercury much as chalk will water. The term *pore*, in physics, is restricted to the invisible spaces that separate molecules. The cavities that may be seen in a sponge are not pores but holes; they are no more entitled to be called pores than the cells of a honeycomb or the rooms of a house are entitled to be called, respectively, the pores of the honeycomb or of the house.

Small as animalcules are, they are coarse lumps in comparison with the size of the molecule. By means of delicate calculations, the physicist has succeeded in ascertaining approximately the probable size of the molecule. If a drop of water could be magnified to the size of the earth, it is thought that its molecules would appear smaller than an apple. In other words, the molecule, in size, is to a drop of water what an apple is to the earth. If we should attempt to count the number of molecules in a pin's head, counting at the rate of ten million in a second, we should require 250,000 years.

§ 7. **Density.**—Cut several blocks of wood, apple, putty, lead, etc., of just the same size, and weigh them. Do they have the same weight? Can you explain the difference by a difference of porosity?

Again, if you can make the experiment illustrated in Figs. 4 and 5, using various gases, you will find that the weights of the same volumes of different gases are different. But the chemist has reasons for believing that there is the same number of molecules in the globe whatever be the gas, if the pressure and the temperature are the same. We see then that some bodies have more matter in a given volume than others, either because the molecules are closer together, or because the molecules are different; we call them more *dense*. By the *mass of a body* we understand *the quantity of matter in it*; and by its *density*, the *mass in the unit volume of it*. For example, the density of cast-iron is about 450 pounds per cubic foot.

§ 8. **Simple and Compound Substances.**—Place a small quantity of sugar on a hot stove and hold a *cold* sheet of glass or of polished metal a few inches above it. In a few minutes you will find a black mass of charcoal or carbon on the stove, and some water on the surface of the glass. Evidently the sugar must have contained the carbon and the water. The heat, in your experiment, expels the water in the form of steam, and leaves the carbon. Carbon can be extracted from sugar in another way. Prepare a very thick syrup, by dissolving sugar in hot water, and pour upon the syrup two or three times its bulk of sulphuric acid. You will quickly obtain a bulky, spongy mass of carbon. In this case the water has been absorbed by the sulphuric acid.

By suitable processes, there may be obtained from marble three substances, each one of which is entirely unlike marble. One of the substances is carbon; another is a metal called calcium, which looks very much like silver; the third is a gas called oxygen, which, when set free from its prison-house in the solid, expands to many times the size of the marble from which it was liberated.

If we should grind a small piece of marble for many hours in a mortar, we should reduce the marble to a very fine powder, but should fall very far short of reducing it to its molecules. Still, each little particle of the powder is as truly marble as the original lump. If we should continue the division, in our imaginations, till the marble were reduced to molecules, we should expect to find all the molecules just alike. Now, since our smallest piece, our molecule, our unit of marble, is *marble*, and since marble is composed of the three substances, carbon, calcium, and oxygen, we conclude that our molecule itself must be capable of division. No one has been able to separate any one of these substances into other substances. No one has taken away from calcium anything but calcium, or extracted from carbon or from oxygen anything but carbon or oxygen.

Those substances that have resisted all efforts to break them up into other substances are called *simple substances* or *elements*. Those substances that may be broken up into other substances are called *compound substances*. Of the large number of substances known to man, only about 70 are elements. All other substances are compounds of two or more of these elements.

A molecule of any substance, simple or compound, is that minute mass of the substance which cannot be divided without destroying its properties.

§ 9. Physical and Chemical Changes.

—When sugar is ground to a powder, the particles are simply torn apart, but do not lose their characteristics. The powder is just as sweet as the lump. Such a division is called a *physical division*. Generally, *any change in a substance that does not cause it to lose its identity, in other words, to cease to be that substance, is called a physical change.* When sufficient heat is applied to sugar, the molecules themselves are divided; and when a molecule of sugar is divided, the result is not two parts of a molecule of sugar, but the two substances, carbon and water. The sweetness is destroyed; sugar no longer exists; other substances have taken its place. The molecule of sugar is no more like the substances into which it has been separated, than a word, as a whole, is like the letters that compose it. Such a division is called a *chemical division*. Generally, *any change in a substance that causes it to lose its identity, or cease to be that substance, is called a chemical change.*

Ice, heated, melts to water; water, heated, becomes steam; steam, cooled, condenses to water; water, cooled, becomes solid. During these changes, the substance (the molecule) has not changed. There has been only a change *among* the molecules, in distance and arrangement. What kind of change is this? But if the steam is subjected to a very intense heat, the result is that it becomes converted into a mixed gas, consisting of two gases, oxygen and hydrogen. This gas is not condensable at any ordinary temperature. Unlike steam, it burns and even explodes. What kind of separation is this? What has been separated? What has been divided?

Blackboard crayons are prepared by subjecting the dust of plaster of Paris to great pressure, which causes the particles to unite and form the crayon. What kind of change is this? What kind of union? In the experiment (§ 2) with the ammonia and hydrochloric-acid gases, the two gases disappear, and a solid is left in their place. What kind of change is this: chemical or physical? Is it union or separation?

§ 10. Annihilation and Creation of Matter impossible.—Experiment 1.—Prepare a saturated solution of calcium chloride. Mix with an equal bulk of water and weigh the solution. Prepare a dilute solution of sulphuric acid (1 to 4), and pour an equal weight of the last solution on the first, all at once, and shake gently. Instantly the mixed liquid becomes a solid. The solid formed is commonly called plaster of Paris. It is an entirely different substance from either of the two liquids used. What kind of change is this? A new substance has been formed. Has matter been created? Weigh the resulting solid; compare its weight with the sum of the weights of the two liquids. What do you find? What conclusion do you draw?

Solids may be converted into liquids or gases; gases may be converted into liquids or solids; substances may completely lose their characteristics: but *man has not discovered the means by which a single molecule of matter can be created out of nothing, or by which a single molecule of matter can be reduced to nothing.* Matter cannot be created, cannot be annihilated; it is a constant quantity. The discovery of this fact laid the foundation of the science of Chemistry.

This statement may not seem to accord with many occurrences of everyday experience. Wood, coal, and other substances burn; matter disappears, and very little is left that can be seen. But does matter pass out of existence when it disappears in burning, or does it assume the invisible state known by the name of gas?

Experiment 2.—Hold a cold, dry glass tumbler over a candle-flame. The bright glass instantly becomes dimmed;

and, on close examination, you find the glass bedewed with fine drops of a liquid. This liquid is water.

You may think it strange that water is formed in the hot flame; yet this simple experiment shows that this is really the case. If water is formed during the burning, what is the reason we do not see it? Simply because it rises in the form of steam, which is an invisible gas. The visible cloud, often called steam, which is formed in front of the nozzle of a tea-kettle, is not steam, but fine drops of water floating in the air, —a sort of water-dust. All clouds are of the same nature. A cloud always stands over Niagara Falls, even on the clearest days. The water of the river falls a distance of 150 feet, and some of it is dashed into fragments, or dust, which rise in a cloud.

Experiment 3.—Introduce a candle-flame into a clean glass bottle; after it has burned a few minutes the flame goes out. Why does it go out? See whether the air in the bottle is the same as it was before. Pour a wineglassful of lime-water into the bottle, cover tightly, and shake. Also pour lime-water into a bottle filled with air. What difference do you observe in the results? Does the experiment show that any new substance has been formed during the burning? If so, is it a visible substance? Can you depend upon the sense of sight alone to discover the presence of matter?

Before we can decide whether or not matter is annihilated while burning, it is necessary to collect carefully, not only the ashes, but all the invisible gases that are formed. This is a somewhat troublesome experiment; but it has been frequently performed, and it is found that their collective weight is equal to the weight of the body burned plus the weight of that portion of the atmosphere which disappears in the burning.

Water does not pass out of existence when it "dries up"; nor are raindrops and dewdrops created out of nothing. Matter is everywhere undergoing great and various changes, both chemical and physical. Nature is ever arraying herself in new forms. The sun warms the tropical ocean, converting the liquid into vapour; the vapour rises in the air, is recondensed on mountain heights, and returns in rivers to the ocean whence it came. Geology teaches us that continents and oceans, and even the "everlasting hills," have a birth and decay, as well as whole tribes of animals and vegetables. Although we may be counted among the living ten years hence, our bodies will, ere that, have crumbled into dust; and the matter that will then compose our bodies is to-day to be found mainly in the earth upon which we tread. Change is stamped upon all matter; nothing is exempt. Only the *quantity* of matter remains unchanged.

§ 11. Force.—Experiment 1.— From a piece of cardboard suspend, by means of silk threads, six pith-balls, so that they may be about $2^{cm\,1}$ apart. Procure a clean, dry glass tube, about 40^{cm} long and 3^{cm} in diameter. Rub a por-

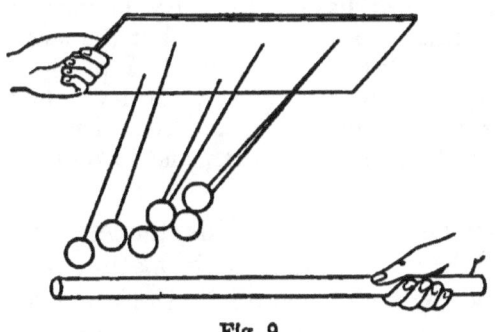

Fig. 9.

tion of this tube briskly with a silk handkerchief, and hold it about 2^{cm} below the balls. The balls seem to become suddenly possessed of life. They gather about the rod, and strive to reach it. If we cut one of the threads, the ball will fly straight to the rod, and cling to it for a time. The means by which the rod acts upon the balls is invisible. Yet evidence is positive that the rod has an influence on the balls. Slip a piece of glass between the rod and the balls; still the influence is felt by the balls. The glass does not sever the invisible bonds that connect the balls with the rod.

Now slowly bring the rod near the balls, till they touch. They at first cling to the rod; but soon the rod, as if displeased with their company, appears to push them away. Withdraw the rod; the balls do not hang by parallel threads as before, but appear to be pushing one another apart. Gradually bring the palm of the hand up beneath the balls, but without touching them. The balls gradually yield to the influence of the hand, and come together. Remove the hand, and they again fly apart. Matter seems not to be the dead, inert thing which it is often called; it can *push* and *pull*.

Experiment 2.—Raise one of these balls with the fingers, and then withdraw the fingers. Something from below seems to reach up, and pull the ball down again. The same happens with each one of the balls; every ball moves as if pulled by something below. What is it that moves the balls? Carry the balls into another room, the same thing occurs. Carry them to any part of the earth, the same thing occurs. It must be on account of the earth itself that the balls have a tendency to fall. All objects have a tendency to move towards the earth. Why this is so we know not, but the fact is obvious.

Attempt to break a string, or crush a piece of chalk, and you find that, notwithstanding, as we believe, the molecules of these bodies do not touch one another, there exists a force which tends to keep them together, and to resist your attempt to separate them.

§ 12. Force defined.—This tendency to push and to pull, which is associated with matter, is called *force*. We do not know why separate portions of matter tend to approach one another, or to separate from one another. Indeed no one has as yet succeeded even in imagining a reasonable cause. We do not know the nature of force; we cannot see it or grasp it; we simply know that there must be a *cause* for certain *effects* produced. The familiar effects produced are motion and rest. For example, we see a body, previously at rest, move; we know that there is a cause: that cause we call force. When a body in motion comes to rest, we look for a cause, and that cause we call force. It is difficult to define force; probably the most comprehensive definition that has been given is the following: *Force is that which can produce, change, or destroy motion.*

All force exhibits itself in pushes or pulls. All change of motion is produced by pushes or pulls, or by a combination of both. A pulling force is called an *attractive* force, or simply *attraction*. A pushing force is called a *repellent* force, or *repulsion*.

§ 13. **Attraction or Repulsion mutual.** — **Experiment.** — Suspend a wooden lath in a sling. Rub one end of a glass rod with silk, and bring that end of the rod near to one end of the lath. Now place the rod in a sling, and bring the lath near to its excited end. Does the experiment prove that the pulling force acts upon only one, or upon both of the bodies? In the experiment with the pith-balls (§ 11, Exp.) they seem to be mutually pushing each other.

All attractions and all repulsions between different portions of matter are mutual.

§ 14. **Molar and Molecular Forces.** — The glass rod does not seem to possess any particular influence until it is rubbed with the handkerchief. The pith-balls do not repel one another until they have first touched the glass rod. After a time, the rod and the balls lose both their powers to attract and to repel. Or, if we pass the hand several times over the part of the rod that has been rubbed, and over the balls, they quickly surrender their special powers. These powers are temporary. They are called *electric powers*, and their cause, whatever it may be, is called *electricity*. The attractive force that draws the balls to the earth existed before the experiment. No manipulation can destroy it or increase it; it is eternal and unchangeable, and exists between all portions of matter. This force is called the *force of gravity*, and the phenomenon is called *gravitation*.

We have seen the effects of attractive and repellent forces, acting across sensible distances. Have we any evidence that these forces exist among portions of matter, at insensible distances, *i.e.* at distances too short to be perceived by our senses? Stretch a piece of india-rubber; you observe that there is a force resisting you. You reason that if the supposition be true, that the grains or molecules that compose the piece of india-rubber do not touch each other, then there must be a powerful attractive force acting across the spaces between the molecules, to prevent their separation. After stretching the india-rubber, let go one end. It springs back to its original form. What is the cause? Compress the india-rubber; its volume is diminished. (Does this confirm our supposition respecting the granular structure of matter?) Remove the pressure; the india-rubber springs back to its original form. What is the cause?

Every body of matter, with the possible exception of the molecule, whether solid, liquid, or gaseous, may be forced into a smaller volume by pressure,—in other words, *matter is compressible.* When pressure is removed, the body expands into nearly or quite its original volume. This shows two things: first, that *the matter of which a body is formed does not really fill all the space which the body appears to occupy;* and, second, that *in the body is a cause, which, acting from within outward, resists outward pressure tending to compress it, and expands the body to its original volume when pressure is removed.* What portion of the theory of the constitution of matter (§ 5) will account for this expansion of a body after a brief compressing force is removed?

It will be seen that the foregoing phenomena may be satisfactorily explained by supposing that the molecules of the india-rubber have a tendency to rush together, or, as it is usually expressed, have a mutual attraction for one another, while at the same time they are in rapid motion beating against one another and driving one another back. What we call *heat* is believed to be this molecular motion. Hence the hotter a body is, the greater its volume, and the more easily is one part of it separated from another part. Now we have the key to the solution of a difficulty, which always arises in the mind of a beginner in science, when he first hears the startling statement that the molecules of bodies, of his own body even, do not touch one another. If faith were of quick growth, he would shudder at the thought of falling to pieces, or of being wafted away by the winds as so much dust.

The ancients, finding it necessary (in order to explain phenomena with which they were familiar) to suppose that matter must be built up of small parts, overcame this difficulty by supposing that the minute particles have hooks or claws by which they grasp one another. Our knowledge of the operation of forces enables us to dispense with hooks and claws, much to the advantage of science. We see that the molecules of a body are kept from falling apart, or from separation, by a universal attractive force; they are also kept from permanent contact by an ever-existing motion among themselves (heat). These forces act at insensible distances between molecules, and hence are called *molecular forces*. When forces act between bodies at sensible distances they are called *molar forces*. Give illustrations (1) of molar forces; (2) of molecular forces.

§ 15. **Matter** presents itself in three different states: *solid*, *liquid*, and *gaseous*,—fairly represented by earth, water, and air. Because these forms are so common and abundant, some ancient philosophers held that all solid matter is formed of earth, all liquids of water, and all gases of air. On this account they called them, together with fire, elements or primary matter. They cannot now be so regarded from a chemical point of view, because each of them has been separated into still more simple substances; nor from a physical standpoint, because, as will soon be shown, many substances may exist in any one of these states.

§ 16. **Characteristics of each of these States.**—**Experiment 1.**—Provide two vessels, a cubical dish and a goblet, each having a capacity of about 200^{ccm}. Also provide 200^{ccm} of sand, 200^{ccm} of water, and a cubical block of wood containing 200^{ccm}. Grasp the block, and place it in the cubical vessel. Attempt to do the same thing with the water. Why can you not grasp the water? Pour a portion of the water into the cubical vessel. When you move a portion of the block, the whole block moves. When you pour a portion of the water into the cubical vessel, the whole does not necessarily go. Why is this? Why is it that we can dip a cupful of water out of a pailful, without raising the whole? Pour all the water into the goblet. The water adapts itself to the shape of the goblet, and the vessel is filled. Attempt to place the block of wood in the goblet. What difference in phenomena do you observe? Why this difference? Pour the sand from vessel to vessel. It adapts itself to the shape of each vessel. Why? Drop the block of wood on a table. Pour water on the table. How does a liquid behave when there is no vessel to confine it?

Experiment 2.—Throw small particles of sawdust into the goblet of water; you can thus render perceptible any motion of the water in the goblet, just as, by throwing blocks of wood on the smooth surface of a river, you can discover the motion of the river. Notice the ease with which the particles move about, rise, and sink. As they become quiet, slightly jar the vessel, or tap it with the end of a pencil, and notice the ease with which disturbance is produced throughout the liquid. Now rap the side of the block with a hammer, and observe how immovable are the particles of wood.

Our experiments teach us that *the molecules of solids are not easily moved out of their places;* consequently, *solid masses form such a firmly connected whole that their shape is not easily changed, and a movement of one part causes a movement of the whole.* On the other hand, *the molecules of liquids have scarcely any fixedness of position, but easily slip among and around one another;* consequently, liquid bodies easily mould themselves to the shape of the vessel that contains them, are *poured* from vessel to vessel, and are easily separated into parts.

But what shall we say of the sand, which, like water, adapts itself to the shape of the containing vessel, and can be poured? Is sand a liquid? and are powders liquids? No, powders are a collection of small *lumps* of solid matter. When powders are poured, lumps of matter roll around one another, as when potatoes are poured from basket to basket. When liquids are poured, *molecules* glide past one another.

It is not so easy to study the characteristics of gases, because we cannot, usually, see them. But we may be aided by a device similar to that employed to make the movement of water visible.

Experiment 3.—Darken a room, and admit, through a small crack or hole, a beam of direct sunlight. You see particles of dust dancing in the path of the light; the motion never ceases. See how easily the motion is quickened by gently waving the hand at some distance from the beam of light.

Experiment 4.—Place under the receiver of an air-pump a partially inflated balloon (Fig. 83), and exhaust the air. The tendency of gases to expand becomes evident.

In gases, fixedness of position of the molecules is entirely wanting, and freedom of motion among themselves is almost perfect. They appear to be in a continual state of repulsion, and consequently gases have a tendency to expand to greater and greater volumes. They expand indefinitely, unless confined by pressure, while liquids and solids tend to preserve a uniformity of volume.

Liquids do not rise above what is called their surface, except as they evaporate, that is, change to gases, and we may have a vessel half full of a liquid; but *gases have no definite free surface*, and there is no such thing as a vessel half full of gas and otherwise empty. On the other hand, *if gases are subjected to pressure, their volume may be very greatly diminished;* for instance, the air that now fills a quart vessel may be compressed into a pint vessel, or even into less space, if sufficient force is used. *The compression of liquids is barely perceptible, even when the pressure is very great.*

§ 17. Philosophy of the three States of Matter. —We conclude from the difficulty which we experience in separating the parts of a solid body, that the molecular attractive force in solids is very great. From the ease with which we usually separate the parts of a body of liquid, we might conclude that this force in liquids is very weak. But, before arriving at any conclusion, it is necessary to consider how the difficulty of separation of the parts of a liquid is to be measured. It is very easy to tear off a portion of a sheet of tinfoil, but we should not surely regard this as an evidence that the molecules of tin have but little attraction for one another, for in tearing such a body we apply the force only to comparatively few molecules at a time. We can form a just estimate of the strength of molecular attraction only by attempting to separate the foil into two portions by such means as that the separation may take place no sooner at one point than at another. So, too, it is very easy to separate a drop of water into two portions, but this is no measure of the attractive forces among its molecules, unless we take precautions that we do not apply the separating force successively to different molecules. If we succeed in preventing such a successive action, and there are certain methods of doing this more or less perfectly, we shall find the process much more difficult,—more so, indeed, than to produce a similar change in many solids.[1]

There is, however, a difference in the molecular action in solids and in liquids; such that, in the latter state, the molecular forces offer no resistance to a *shaping* force, while in the former state, change of shape can only be brought about by the application of considerable force.

In a gas, on the contrary, there is little attraction between the molecules; but as they are constantly hitting one another, and thereby tending to drive one another apart, it requires an external force to keep them together.

NOTE.—In gases the molecules are thought to be in motion like gnats in the air; in liquids, like men moving through a crowd; in solids, the motion of each molecule is like that of a

---

[1] The cohesive force of water is at least 132 lbs. per square inch.—MAXWELL.

man in a dense crowd, where it is almost or quite impossible to leave his neighbours, yet he may turn around, and have some motion from side to side.

Practically, the condition of any portion of matter depends upon its temperature and pressure (see § 128). Just as at ordinary pressures water is a solid, a liquid, or a gas, according to its temperature, so any substance may be made to assume any one of these forms unless a change of temperature occasions a chemical change.

There are certain apparent exceptions to the last statement; for example, charcoal, though it has been vaporised, has never been obtained in a liquid state, doubtless because sufficient pressure has never been used. Ice will change to a vapour, but cannot be melted unless the pressure exceeds six grams per square centimeter. At ordinary pressure, iodine and camphor vaporise, but do not melt. Only within the last few years have physicists been able to produce a temperature low enough to solidify alcohol, and a temperature low enough, combined with a pressure great enough, to liquefy such gases as oxygen, nitrogen, and hydrogen.

As regards the temperature at which different substances assume the different states, there is great diversity. Oxygen and nitrogen gases, or air,—which is a mixture of the two,—have been liquefied and solidified only at extremely low temperatures; and then, only when the intermolecular attraction is aided by tremendous pressure. On the other hand, certain substances, as quartz and lime, are liquefied only by the most intense heat generated by an electric current. The facts, summed up, are as follows: *no one of the three states of matter, solid, liquid, or gaseous, is peculiar to any substance;*[1] *the state that a substance*

[1] We probably should except those compounds which, so far as we know, undergo chemical change without changing to liquids or vapours. But even in the case of these it is not impossible that any one might be liquefied or vaporised without chemical change could we secure the necessary *pressure* as well as the necessary temperature.

assumes depends solely on its temperature and pressure; so that every solid may be regarded as simply matter in a frozen state, every liquid as matter in a melted state, and every gas as matter in a state of vapour. Every liquid has been solidified and volatilised, and every gas has been liquefied and solidified. Air was one of the last of the gases to surrender its reputation of being a "permanent gas." Not till the year 1878 was it reduced to lumps. We may predict the future of our globe. If its heat increases sufficiently, the whole world will become a thin gas. If its heat diminishes indefinitely, that is, until *all* molecular motion ceases, all earth and air will become a solid mass.

PHENOMENA OF ATTRACTION

According to the circumstances under which attraction acts, we have the various phenomena called *gravitation, cohesion, adhesion, capillarity, chemism,* and *magnetism*. Sometimes these terms are used as names of the unknown forces that cause the phenomena.

§ 18. **Gravitation.**—That attraction which exists between all matter, at all distances, is called *gravitation*. Gravitation is universal, that is, every molecule of matter attracts every other molecule of matter in the universe. The whole force with which two bodies attract each other is the sum of the mutual attractions of their molecules, and depends upon the number of molecules the two bodies individually contain, and the mass of each molecule. The whole attraction between an apple and the earth is equal to the sum of the attractions between every molecule in the apple and every molecule in the earth.

§ 19. **Weight.**—It is scarcely necessary to state, that what is understood by the *weight* of a body is the pressure it exerts in a vertical direction, and is (approximately at least) the mutual attraction between it and the earth.

The term *mass* is equivalent to the expression *quantity of matter*. We assume that, at the same point on the surface of the earth, *weight* is proportional to *mass*. Why do we weigh articles of trade, such as sugar and tea ?

§ 20. Does the Apple attract the Earth with as much Force as the Earth attracts the Apple ? —Let us examine this question. First assume that the molecules of the apple and the earth have equal masses, *i.e.* are homogeneous ; then it is evident that the attraction of any molecule in the apple for any molecule in the earth is equal to the attraction of any molecule in the earth for any molecule in the apple. That is, if the earth and the apple consisted each of a single like molecule, their attractions for each other would be equal. Now, suppose that the apple contains two and the earth five such molecules. Let the force with which one molecule attracts another be represented by n. Now if each molecule of the apple attracts the five molecules in the earth with a force of $5\,n$; the two molecules in the apple would attract the earth with a force of $10\,n$. On the other hand, each molecule of the earth attracts the two molecules of the apple with a force of $2\,n$, and the five molecules in the earth would attract the apple with a force of $10\,n$. It is obvious that the same course of reasoning will apply in case the attraction is between two molecules whose masses differ, and consequently between all bodies of whatever mass or substance. Hence does it appear that a body of small mass attracts a body of large mass as strongly as the latter attracts the former ?

If the apple attracts the earth as strongly as the earth attracts the apple, why does not the earth rise to meet the apple ? Let us examine a similar case. Suppose that a man in a boat pulls on a rope attached to a ship. His pulling draws the boat to the ship ; but the ship does not appear to move. But if five hundred men, in as many boats, pulled together, the ship would be seen to move. Did one man produce no motion ? If so, then would the five hundred men produce no motion, since five hundred times nothing is nothing.

You will learn, in the next chapter, that the space through which a given force moves a free body in a given time varies inversely as the mass of the body. Does this fact explain the foregoing phenomena ?

§ 21. The Force of Gravity varies with the Distance from the Centre.—Observations made in various ways show that the force of gravity varies over the surface of the earth. It can be proved by geometrical methods that a sphere, composed of homogeneous spherical shells, acts upon a molecule without it as though all its attractive force were concentrated at its centre. Now it is found that the nearer an object *without the earth's surface* is to the centre of the earth the greater is the force of gravity. The polar diameter of the earth is about twenty-six miles less than its equatorial diameter, and, consequently, the distance from the centre to the surface at the poles is thirteen miles less than to the surface at the equator. This considerable difference in distance from the centre occasions an appreciable difference between the weight of a body (having any considerable mass) at the equator and its weight at the poles; and, since the distance of the surface from the centre constantly increases as we go from the poles towards the equator, the weight of all objects transported from the poles toward the equator constantly diminishes.

You will now see one reason why the statement that "weight is proportional to mass" (§ 19) must be restricted to a comparison of *masses at the same place and at the same altitude* only. The propriety of making a distinction between the terms *mass* and *weight* is also apparent, as that which is called mass does not change when a body is transferred from place to place, while that which is called weight may change.

It is obvious that any object raised above the earth's surface, as in a balloon, must weigh less than at the surface of the earth. But the heights with which we commonly have to deal in our experiments are so small in comparison with the earth's radius, that the differences in weight due to differences in height at a given place can scarcely be detected by most delicate tests.

It may be proved that if the earth were of uniform

density, bodies carried below its surface would lose in weight as the distance below the surface increases. At one-fourth the distance to the centre there would be a loss of one-fourth the weight. At one-half the distance the weight would be one-half; and at the centre nothing. Is weight an essential property of matter? State certain conditions on which a body would have no weight.

The terms *up* and *down* are derived from the attraction between the earth and terrestrial objects. *Down* is the direction in which a body falls or tends to move in consequence of gravitation, and is (at least approximately) towards the *centre* of the earth. *Up* is the opposite direction. It is apparent that the up and down of one place cannot be the same direction as the up and down of any other place.

It may be experimentally demonstrated that: *The intensity of the attraction of gravity between two bodies, each of which lies without the other, varies directly as the product of the measures of their masses, and inversely as the square of the distance between their centres of gravity.* This may be expressed thus: Let m denote the measure of the mass of A and n the measure, in terms of the same unit, of the mass of B, and let s denote the measure of the distance between the centre of gravity of A and the centre of gravity of B; then $\frac{m\,n}{s^2}$ is a measure of the attraction of gravity between A and B.

The foregoing statement of fact is called the LAW OF GRAVITATION. This law was discovered and demonstrated by Sir Isaac Newton, and a knowledge of it and of the principles of motion enables us to understand, among other things, the movements of the earth, the moon, and the planets.

QUESTIONS

1. If the earth's mass were doubled without any change of volume, how would it affect your weight?

2. On what principle do you determine that the mass of one body is ten times the mass of another body?

3. To what extent must you increase the distance between the centres of gravity of two bodies that their attraction may become one-fourth of what it was at first?

4. If a body on the surface of the earth is 4000 miles from the centre of gravity of the earth, and weighs at this place 100 pounds, what would the same body weigh if it were taken 4000 miles above the earth's surface, it being weighed with a spring-balance in both cases?

5. The masses of the planets Mercury, Venus, Earth, and Mars are respectively very nearly as 7, 79, 100, and 12; assuming that the distance between the centres of the first two is the same as the distance between the centres of the last two, how would the attraction between the first two compare with the attraction between the last two?

6. What would be the answer to the last question if the distance between the centres of the first two were four times the distance between the centres of the last two?

7. Would the weight of a soldier's knapsack be sensibly less if it were carried on the top of his rifle?

8. If an iron pound-weight and a pound of sugar were balanced with ordinary scales at the equator, and transported to one of the poles of the earth, would they cease to balance each other?

9. If the same quantity of sugar be suspended from a spring-balance at the pole and at the equator, will this instrument indicate just a pound in both cases? If not, in which case will it indicate the most?

10. Imagine yourself at the centre of the earth. In what direction must you turn your face in order to look up?

11. Imagine a person at one of the poles, and another at the equator, to be looking down. Would they both look in the same direction?

12. Draw a circle to represent the earth, and a line to represent the direction in which each person would look.

13. To what is "water power" due?

14. To what are the tides due?

15. Which is more difficult, to ascend or to descend a hill, and why?

16. The earth has about 81 times as much matter in it as the moon, and its diameter is about 4 times that of the moon.

At which body would you weigh more? Applying the "law of gravitation," compare your weight at the surface of the earth with your weight at the surface of the moon.

17. Is there a place between the two bodies at which you would weigh nothing? If so, why? Where is it?

18. How far does the earth's attraction extend?

19. Which would you prefer, a pound of gold weighed with a spring-balance at the surface of the earth, or a pound similarly weighed 3,000,000^m below the surface?

§ 22. Cohesion.—That attraction which holds the molecules of the same substance together, so as to form larger bodies, is called *cohesion*. It is the force that prevents our bodies, and all bodies, from falling down into a mass of dust. It is that force which resists a force tending to break or crush a body. It is greatest in solids, usually less in liquids, and very weak in those substances which are gases at ordinary temperature and pressure. Its effect is noticeable only at insensible distances, and it is strictly a molecular force. When once the cohesion is overcome, so as to separate the molecules even a very little, it is difficult to force them near enough to one another for this force to become effective again. Broken pieces of glass and crockery cannot be so nicely readjusted that they will hold together. Yet two polished surfaces of glass, placed in contact, will cohere quite strongly. Or if the glass is heated till it is soft, or in a semi-fluid condition, then, by pressure, the molecules at the two surfaces may be made to flow around one another and pack themselves closely together, and the two bodies will become firmly united. This process is called *welding*. In this manner iron is welded.

Cohesive force varies greatly in different substances, probably according to the variation in the nature, form, and arrangement of the molecules of which they are composed. These modifications of the force of attraction of cohesion give rise to certain *conditions of matter*, designated as *crystalline, amorphous, hard, flexible, elastic, brittle, viscous, malleable, ductile,* and *tenacious*.

§ 23. Crystalline and Amorphous Conditions of Matter.—If our vision could be rendered keen enough to enable us to see and examine the molecular structure of different substances, to look into their bodies, as we look into the starry heavens, and observe the positions, the spaces, and the arrangement of that unexplored world, there would undoubtedly be unfolded to us wonders and beauties of which we have never dreamed. We should probably behold an endless variety of arrangement among the molecules. We might learn why it is that although the molecule of the diamond, of graphite, and of charcoal is the same (*i.e.* the same substance), we get, possibly by different arrangement and different behaviour of molecular forces, the hard, transparent, and brilliant diamond in the one case, the soft, opaque, metallic-looking graphite in another, and finally the porous, black, and shapeless charcoal.

Obtain a piece of mica and a piece of chalk, and attempt to cut them in two, by applying the knife in different directions. You find that you can easily cleave the mica in one direction, and obtain a smooth, shining surface. This is called its *plane of cleavage*. Cut it in any other direction, and you get rough and ragged surfaces. The chalk may be cleft in one direction as well as another, and in no direction can a smooth surface be obtained. We learn by these trials that matter may have method in its arrangement, or possess definite structure.

When matter exhibits structure or method in its molecular arrangement, it is said to be *crystalline*. Examples of crystalline arrangement are mica, Iceland spar, and carbon in the form of diamond. When the molecular arrangement of matter is methodless or structureless, the matter is said to be *amorphous*. Examples of amorphous matter are chalk, glue, glass, and carbon in the form of charcoal.

Experiment 1.—Pulverise 20^g of alum, and dissolve in 50^{ccm} of hot water; suspend a thread in the solution, and put it away where it can quietly and slowly cool. The process by which matter, in solidifying, assumes a structural condition is called *crystallisation*, and bodies which have acquired regular shape by this process are called *crystals*. Obtain crystals of saltpetre, of blue vitriol, and of potassium bichromate, by dissolving as much as possible of these substances in hot water, and allowing the solutions to cool, *always slowly and quietly*.

Experiment 2.—Thoroughly clean a piece of window-glass, and pour upon it a hot concentrated solution (see § 36) of ammonium chloride or of saltpetre. Allow the liquid to drain off, and hold it up to the sunlight. What do you observe? What is it that is going on before your eyes?

Very interesting illustrations of crystallisation are those delicate lacelike figures which follow the touch of frost on the window-pane. Fig. 10 represents a few of more than a thousand forms of snowflakes that have been discovered, resulting, we may readily believe, from a variety of arrangement of the water molecules.

Nature teems with crystals. Nearly every kind of matter, in passing from the liquid state (whether molten or in solution) to the solid state, tends to assume symmetrical forms. *Crystallisation is the rule; amorphism, the exception.* Break open a sugar-loaf, and you will find the surface fracture composed of small, shining, crystalline surfaces. You can scarcely pick up a stone and break it, without finding crystalline fracture. Every piece of ice is a mass of crystals, so closely packed together that the individuals are not distinguishable.

§ 24. Change of Volume by Crystallisation.—This tendency of matter to structural arrangement is not only very interesting, but very important in the arts. It is very natural to suppose that the new arrangement of molecules, when passing from the liquid to the solid state, should occasion a change in volume. We are able to understand how it is possible that water, in freezing, disregards the law of contraction by cold, and that the molecules are not found so closely packed together, in the new and structural state, as when mingled without any systematic arrangement.

The force exerted by the molecules in changing positions is so enormous as to burst the strongest vessels.

Fig. 10.

Hence our service-pipes are burst when water is allowed to freeze in them. Huge rocks are dislodged from their resting-places in the native quarry on the mountain-side by water getting into the crevices, freezing, expanding year after year, and pushing the rocks from their support. Cast-iron and many alloys, such as type-metal, expand on solidifying. Such metals may be cast in moulds, since, in expanding, they fill all the minute cavities of the mould. Most metals contract on solidifying. Hence gold, silver, and copper coins require to be stamped. Cast-iron, when broken, exhibits a crystalline fracture. Wrought-iron, when subjected to long-continued jarring, —for instance, in the axles of carriage-wheels and iron cannon,—becomes very brittle, and, when broken, exhibits a very marked crystalline fracture which it would not have shown before long use. It is probable that the molecules of iron, when shaken up by the jarring, are free to arrange themselves in their peculiar method, and that, in this new arrangement, the iron is more subject to cleavage along some lines than when in an amorphous condition.

§ 25. What is the Cause of this almost Universal Tendency of Matter to Crystallise?—We have no absolute knowledge of the doings in the molecular world. But we have very satisfactory methods of judging. Analogy is the light by which we must frequently explore inaccessible space. We determine the laws that govern large, tangible masses, and from these we infer the laws that govern small, intangible bodies, such as molecules. Let us adopt this method in attempting to unravel the mystery before us.

Experiment 1.—Take two cambric needles, and draw each several times, from the eye to the point, over the same end of a magnet. Now suspend each needle by a thread, so that it will be balanced in a horizontal position. Bring the eye of one near the point of the other. Bring the middle of one near the middle of the other. Bring the point of one near the point of the other. Bring the eye of one near the eye of the other. What is the result in each case? The opposite character which the ends of the needle exhibit we will call *polarity*.

Now, break one of the needles into two pieces, and experiment as before. Break one of these into still smaller pieces, and the smallest piece that you can obtain possesses polarity, as certainly as the original needle. Imagine the work of division to be continued till the molecule is reached. Is it too much to assume that the molecule may possess polarity?

Throw a dozen of these magnetised needles on a sheet of paper, and shake them about. Do they group themselves with any appearance of regularity? If two are to form a rigid group, how must they be arranged? How is it with a group of three?

Experiment 2.—Next place a magnet beneath a sheet of paper, and sift iron filings over it. Gently tap the paper. The result reminds us of the effect of jarring on the carriage-axle and the cannon, where molecules, disturbed tend to arrange themselves according to some guiding principle. Next, lay the magnet on a bed of iron filings, and then raise it. Carefully examine the result.

We pass readily from these facts to conclusions respecting the molecular arrangement in the crystal. Only grant the supposition that the molecule is endowed with something similar to polarity, and we can picture to ourselves the molecules, like the magnetised iron filings, wheeling into line in obedience to their polar forces. Crystals are more easily cleft in some directions than in others; may not this be accounted for by supposing that, like the magnet, the attraction on some sides of the molecule is greater than on others?

§ 26. Hardness.—Name some metal that you can scratch with a finger-nail. See if you can scratch a piece of copper with a piece of lead, and *vice versâ*. Get specimens of as many as possible of the following substances: talc, chalk, glass, quartz, iron, silver, lead, copper, rock-salt, and marble. Ascertain which of them will scratch glass, and which are scratched by glass. What term do we employ in speaking of those substances that are easily scratched? To those that are scratched with difficulty? Which is the softest metal that you have tried? The hardest? Which is the softer metal, iron or lead? Which is the more dense metal? Does hardness depend upon density? What force must be overcome in order to scratch a substance? When will one substance scratch another?

To enable us to express degrees of hardness, the following table of reference is generally adopted :—

MOHR'S SCALE OF HARDNESS

1. Talc.
2. Gypsum (or Rock-Salt).
3. Calcite.
4. Fluor-Spar.
5. Apatite.
6. Orthoclase (Feldspar).
7. Quartz.
8. Topaz.
9. Corundum.
10. Diamond.

By comparing a given substance with the substances in the table, its degree of hardness can be expressed approximately by one of the numbers used in the table. If the hardness of a substance is indicated by the number 4, what would you understand by it?

§ 27. Flexibility.—Such substances as may be bent, or as admit of a hinge-like movement among their molecules, are called *flexible*. What difference have you noticed in different jack-knife blades? How can you tell a soft blade from a hard blade? If you bend

Fig. 11.

a stick, as in Fig. 11, how must the molecules on the upper side be affected? How those on the lower side? Make experiments in bending a long thin glass rod or tube. What conclusions regarding the molecules of glass do you reach?

§ 28. Elasticity.—Obtain thin strips of the following substances: india-rubber, wood, ivory, whalebone, steel, brass, copper, iron, zinc, and lead. Stretch the piece of india-rubber. What change in its molecular condition must occur when it is stretched? What molecular force causes it to contract when the stretching force is removed? Compress the india-rubber. What change of molecular condition takes place in compression? What causes it to expand when the pressure is removed? Bend each one of the above strips. Note which *completely* unbends when the force is removed. Arrange the names of these substances in the order of the rapidity, and also in the order of the completeness with which they unbend.

What change takes place among the molecules on the concave side of the bent strips? What, among the molecules on the convex side? What two forces are concerned in the unbending? Twist the cord of a window-tassel. What causes it to untwist? The property which matter possesses of recovering its former shape and volume, after having yielded to some external force, is called *elasticity*. To what forces is elasticity due? Does all matter possess this property in the same degree? Does the india-rubber possess the same ability to unbend as to contract after being stretched? In what four ways have you tested the elasticity of substances? Does a substance possess equal power of recovering its form after yielding to each of these four methods of applying force? Why are pens made of steel? What moves the machinery of a watch? What is the cause of the softness of a hair mattress or of a feather-bed?

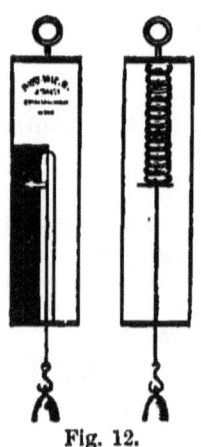
Fig. 12.

A common spring-balance used for weighing consists of a steel spring wound into a coil. The weight of the body to be weighed straightens or draws out the spring. A pointer moving over a plate which is divided into equal parts shows how much the spring has been drawn out. With known weights experiment with the spring-balance to ascertain the relation between the drawing out of the spring and the stress which produces that drawing out. State the result in the form of a "law" (*i.e.* make a *general* statement based upon your several observations). The entire virtue of this apparatus consists in the elasticity of the spring, or its power to recover its original form after being drawn out. Give other illustrations of the application of elasticity to practical purposes.

Any alteration in the form of a body due to the application of a force is called a *strain*, and the force by which the strain is produced is called the *stress*. A body which, having experienced a strain due to a certain stress, completely recovers its original condition when the stress is removed, is said to be *perfectly elastic*. Liquids and gases are perfectly elastic (see § 97). Solids are perfectly elastic up to a certain limit, which varies greatly in different substances. If the strain exceeds a certain limit, the form of the solid becomes permanently altered, and the state of the body, when the permanent alteration is about to take place, is called the *limit of perfect elasticity*. In soft or plastic bodies this limit is soon reached. What is the result of overloading carriage springs?

§ 29. **Brittleness.**— Apply sharp blows with a hammer to each of the substances whose hardness you have tested (§ 26), and ascertain which are the most easily broken or pulverised. Observe that some substances suffer a permanent change in form when subjected to a stress which exceeds their limit of elasticity, while others break before there is any permanent alteration in form. The latter are said to be *brittle*. Name substances which, *from your own observation*, you know to be "brittle."

§ 30. **Viscosity.**—Support in a horizontal position, at one of its extremities, a stick of sealing-wax, and suspend from its free extremity a small weight, and let it remain in this condition several days, or perhaps weeks. At the end of the time the stick will be found permanently bent. Had an attempt been made to bend the stick quickly, it would have been found quite brittle.

A body which, subjected to a stress for a considerable time, suffers a permanent change in form, is said to be *viscous*. Hardness is not opposed to viscosity. A lump of pitch may be quite hard, and yet in the course of time it will flatten itself out by its own weight, and flow down hill like a stream of syrup. Liquids like treacle and honey are said to be viscous, in distinction from limpid liquids like water and alcohol.

§ 31. **Malleability and Ductility.**—Some substances possess, in the solid state, a certain amount of *fluidity*; that is, their molecules may be displaced without overcoming their cohesion. Place a piece of lead on an anvil, and hammer it. It spreads out under the hammer into sheets, without being broken, though it is evident that the molecules have moved about among one another, and assumed entirely different relative positions. Heat a piece of soft glass tube in a gas-flame, and, although the glass does not become a liquid, it behaves very much like a liquid, and can be drawn out into very fine threads. When a solid possesses sufficient fluidity to admit of being drawn out into threads, it is said to be *ductile*.[1] When it will admit of being hammered or rolled into sheets, it is said to be *malleable*.[2]

As might be expected, *those substances that are ductile are also malleable*. But the same substance does not usually possess the two properties in an equal degree. Platinum is the most ductile metal. It can be drawn into wire finer than a spider's thread. It is the seventh metal in the rank of malleability. Gold is the most malleable metal. It can be hammered into leaves so thin, that it would require 300,000 to make a book one inch thick. It ranks next to platinum in ductility. Iron, at a red heat, is very malleable and ductile. What metals can be drawn into wires? What metals can be rolled or hammered into sheets?

§ 32. **Tenacity.**—In order that a substance may be

---

[1] Ductile, *draw-able*. [2] Malleable, as it were *mallet-able*.

ductile, it is evident that there must exist strong cohesion among its molecules, and that this cohesion is not weakened by the change in the mere *arrangement* of the molecules which results from the drawing out. The power that matter possesses of resisting rupture, by a pulling force, is called *tenacity*.[1] *A body may be tenacious without being ductile, but it cannot be ductile without being tenacious.* It is remarkable that the tenacity of most metals is increased by being drawn out into wires. It would seem that, in the new arrangement which the molecules assume, the cohesion is stronger than in the old. Hence cables made of iron wire twisted together, so as to form an iron rope, are stronger than iron chains of equal weight and length, and are much used instead of chains, where great tensile strength is required.

§ 33. **Adhesion.**—Grasp with your finger a piece of gold-leaf, and, honest as you may be, it will stick to your fingers; it will not drop off, it cannot be shaken off, and to attempt to pull it off is to increase the difficulty. Dust and dirt stick to clothing. Thrust your hand into water, and it comes out wet. You can climb a pole, because your hands stick to the pole; but if the pole is greased, climbing is not so easy. We could not pick anything up, or hold anything in our hands, were it not that these things stick to the hands.

Every minute's experience teaches us that not only is there an attractive force between molecules of the same kind of matter, but there is also an attractive force between molecules of unlike matter. That force which causes *unlike* substances to cling together, is called *adhesion*. Is adhesion a molar or a molecular force? How does it differ from cohesion? Why do not gold watches, and other articles of gold jewellery, appear to stick to the fingers? What keeps nails, driven into wood, in their places? What would happen if all adhesion between the

---

[1] Tenacity, property of *holding*.

different parts of the building you are in should be suddenly destroyed? When a liquid sticks to a solid, what term do we usually employ in describing the phenomenon?

Experiment 1.—Shake a small quantity of olive-oil in water, and observe the form assumed by the fragments of oil as they rise through the water. Does this experiment indicate that the adhesion between the oil and water or the cohesion in the oil is the stronger?

Experiment 2.—Suspend a plate of glass, about 80^{cm} square, from one arm of a scale-beam, attaching the threads to the plate with sealing-wax. Balance it, and place a dish of water under the glass, so that its under surface will just touch the surface of the water. Add small weight until the glass leaves the water. Examine the under-side of the glass. Have you separated the glass from the water, or have you torn the water apart? Do you infer from your experiment that the adhesion between the glass and the water, or the cohesion in the water is the greater?

Fig. 13.

Glass is wet by water, but is not wet by mercury. Is there no adhesion between mercury and glass?

Experiment 3.—Substitute mercury for water in the last experiment. Do you find any indication of adhesion? Is it greater or less than that between glass and water?

It is probable that *there is some adhesion between all substances when brought in contact. If a liquid adheres to a solid more firmly than the molecules of the liquid cohere, then will the solid be wet by the liquid.* If a solid is not wet by a liquid, it is not because adhesion is wanting, but because cohesion in the liquid is stronger. That gases adhere to solids is proved by the phenomena of absorption described in § 37.

QUESTIONS

1. Why will not water wet articles that have been greased ?
2. Why is it difficult to lift a board out of water flatwise ?
3. Why does water run down the side of a tumbler when it is inclined, instead of falling vertically ? Suggest some method of preventing it.
4. In what does the value of cement, glue, and mucilage consist ?
5. What property of matter enables you to leave a mark with a pencil or crayon ?

§ 34. **Capillarity.**—Examine the surface of water in a goblet. You find the surface level, as in A (Fig. 14),

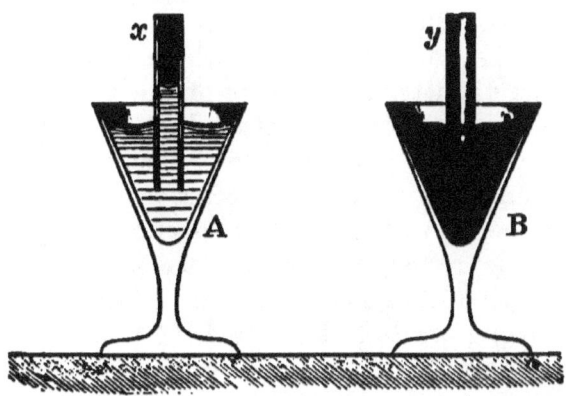

Fig. 14.

except around the edge next the glass, where the water is curved upward so as to resemble the interior surface of a watch glass. Mercury placed in a goblet (B) has its edge turned downward, resembling the exterior surface of a watch glass. This seems to indicate a repulsion between mercury and glass. But a previous experiment (§ 33) has shown that, instead of repulsion, there is a slight adhesion between these substances.

Pour any liquid on a level surface which it does not wet,—*e.g.* water on paraffin or wax, or mercury on glass. It spreads itself over the surface, but the edges are everywhere rounded or turned down like the edges of mercury in a goblet. Surely these rounded edges are not caused by the repulsion of the sides of a vessel. It has been suggested that the edges of all liquids will be turned down unless the adhesion between them and the sides of the vessels exceeds the cohesion in the liquid. The glass does not cause the turning down of the surface of mercury in the goblet,—its tendency is rather to prevent it.

Thrust vertically two plates of glass into water, and gradually bring the surfaces near each other. Soon the water rises between the plates, and rises higher as the plates are brought nearer. Thrust a glass tube of very fine bore into water; the attraction within it, on all sides, will raise the water to twice the height it would reach when between two plates whose distance apart is equal to the diameter of the bore of the tube. Thrust a tube of the same bore into alcohol; this liquid rises in the tube, but not so high as water. The surfaces of both the water and the alcohol are concave. If the tube is placed in mercury, the opposite phenomena occur: the mercury is depressed, and its surface is convex.[1] Both ascension and depression diminish as the temperature increases, being greatest at the freezing point of the given liquid, and least at its boiling point. (What does this fact indicate regarding the relative effects of heat upon cohesion and upon adhesion?) Inasmuch as the phenomena are best shown in tubes having bores of the size of hairs, they are in such cases called *capillary*[2] *phenomena*, and the tubes are called *capillary tubes*.

The phenomena of capillary action are well shown by placing various liquids in U-shaped glass tubes, having one arm reduced to a capillary size, as A and B in Fig.

---

[1] The scope of this book will not admit of an explanation of the phenomena of capillarity. The student can find a lucid treatment of this subject in Maxwell's *Theory of Heat*, pp. 260-274; also under "Capillary Action," *Encyclopædia Britannica*.

[2] Capillary, *hair-like*.

15. Mercury poured into A assumes convex surfaces in both arms, but does not rise so high in the small arm as it stands in the large arm. Pour water into B, and all the phenomena are reversed. C is a glass tube containing water and mercury, and showing the shapes that the surfaces of the two liquids take.

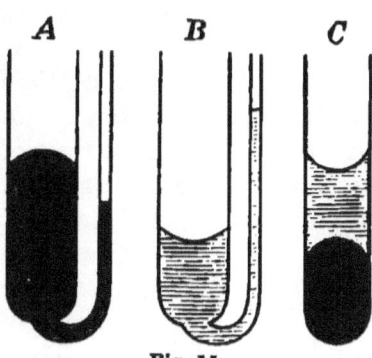

Fig. 15.

Generalising the above facts, we have the *four laws of capillary action* :—

I. *Liquids rise in tubes when they wet them, and are depressed when they do not.*

II. *The ascension or depression varies inversely as*[1] *the diameter of the bore.*

III. *The ascension and depression vary with*[1] *the nature of the substances employed.*

IV. *The ascension or depression varies inversely with the temperature.*

Illustrations of capillary action are abundant. It feeds the lamp-flame with oil. It wets the whole towel, if one end is left for a time in a basin of water. It draws water into wood, and causes it to swell with a force sufficient to split rocks, and to raise large weights. How does a little water in a wooden tub prevent its falling to pieces ?

§ 35. **Other Molecular Phenomena.**—Besides the phenomena we have just studied, there are a great many others depending in part on molecular attraction, but much more on the molecular motions, of which we learned in § 5. Many of them are quite familiar and important ;

---

[1] Observe that throughout this treatise the word *as* expresses an exact proportion. When there is not an exact proportion, the word *with* is used.

but the explanation, even when it can be given, is as yet complicated and incomplete. The principal names given these phenomena are *solution, absorption,* and *diffusion.*

§ 36. **Solution of Solids**—*depends, probably, mainly on molecular attraction.* Hold a lump of sugar so that it will just touch the surface of water. Soon water is drawn up into the pores of the lump by capillary action, and the whole lump, including the part not submerged, becomes moist. Next you discover that the lump becomes smaller, and slowly disappears in the water.

When a solid becomes diffused through a liquid, it is said to be *dissolved.* The dissolving liquid is called a *solvent,* and the resulting liquid is called a *solution.* It has been supposed that *a liquid will dissolve a solid, only when the adhesion between them is greater than the cohesion in the solid.* A liquid always dissolves a solid more rapidly at first, less rapidly as the adhesion becomes more nearly satisfied; and when it is completely satisfied, or is balanced by the cohesion in the solid, the liquid will dissolve no more of the solid, and the solution is said to be *saturated.* When a solution will take much more of a solid, it is said to be *dilute;* and *concentrated,* when it will take little or no more.

If the solid be first pulverised, the liquid has more surface on which to act, and the solid is dissolved much more rapidly. With few exceptions, hot liquids dissolve solids more rapidly and in greater quantities than cold liquids. What does this fact indicate regarding the relative effects of heat upon cohesion and upon adhesion? Compare your answer to this question with your answer to a similar question in § 34. Does it appear that, as yet, the hypotheses regarding capillary attraction and solution are satisfactory? Boiling water dissolves three times as much alum as cold water; consequently, when a hot saturated solution of alum is allowed to cool, at least two-thirds of the alum must be restored to the solid state (see Exp. 1, § 23), while one-third, or the amount that the cold liquid is capable of dissolving, remains in solution. The remaining solution is called the *mother-liquor*. Lime and a few other substances are dissolved better in cold water. Crystals of such substances are only obtained by gradual evaporation of the solvent.

Water is the great solvent. When we speak of the solubility of a substance, water is always understood to be the solvent, unless some other liquid is specified. Why is it fortunate that water is so good a solvent? Name substances that water does not dissolve. Of the many substances insoluble in water, some, as phosphorus, gums, and resin, find a solvent in alcohol; sulphur, in bi-sulphide of carbon; lead, in mercury; and fats, in ether or benzene. Would you wash varnished furniture with alcohol? How are grease-spots removed from clothing?

§ 37. Absorption of Gases by Solids—*depends mainly on molecular attraction, and is generally superficial.* Certain solids possess so strong an attraction for gases that they not only draw the gases into the small cavities or holes within them, but greatly condense them there. It should be carefully noted that the attraction in this case is generally between the gases and the *surfaces of cavities*, and is hence called *superficial*, in distinction from *inter-molecular attraction*, which is the name given to the phenomenon when gases are taken into the *pores* of a body.

Freshly-burned charcoal placed in dry air may, in a few days, have its weight increased one-fiftieth in consequence of the air that it absorbs. (Has air weight?) The attraction of charcoal for noxious gases is especially great, making it very efficient in cleansing the air in hospitals, and in removing noxious odours from putrid animal and vegetable matter by absorbing the foul gases that are generated. It does not check decay, but rather hastens it. A rat, which had been buried in charcoal dust, was uncovered at the end of a month; nothing visible was left but the hair and bones, yet no bad odour was perceptible.

§ 38. Absorption of Gases by Liquids—*depends on molecular attraction and motion, and is intermolecular.* Water, at a temperature of 0° C., is capable of condensing in its *pores* six hundred times its own bulk of ammonia gas.[1] Water thus charged with this gas is called "ammonia water." The amount of gas that a liquid will absorb is increased by pressure. "Soda water" is simply water saturated with carbonic-acid gas under great pressure; it contains no soda. When the pressure is removed a large part of the gas escapes, causing effervescence.

§ 39. Free Diffusion of Liquids—*depends mainly on motion.*—**Experiment 1.**—Into a test-tube containing 20^{ccm} of water, pour about 2^{ccm} of olive-oil, and shake. Do the oil and water mix?

Experiment 2.—Partially fill a glass jar (Fig. 16) with water. Then introduce beneath the surface, by means of a long funnel, a concentrated solution of sulphate of copper. Set the vessel aside for a few days. Does the experiment show any mixing or diffusion in this case?

Fig. 16.

Experiment 3.—Take about 1^{ccm} of bisulphide of carbon, colour it by dropping into it a small particle of iodine, and pour this coloured solution into a test-tube nearly filled with water. Set this tube aside for a week. Do you find any indications of diffusion?

Experiment 4.—Treat glycerine and water as you did the

---

[1] It is possible, and even probable, that in this case the molecules of ammonia gas are not simply diffused among the molecules of water, but there is a *chemical union* between the gas and the water, ammonium hydrate being formed thus: $NH_3 + H_2O = NH_4HO$. The application of heat decomposes the hydrate, producing water and ammonia gas, which latter then escapes. Again, in the so-called solution of carbon dioxide (sometimes called carbonic-acid gas) in water, probably, to some extent, there is a chemical union between the gas and the water, thus: $CO_2 + H_2O = H_2CO_3$.

liquids in Experiment 2. What is the result? In cases where diffusion resulted did the liquid move in currents or by molecules? What conclusions do you draw from these experiments?

If, during the operation of diffusion in the last three experiments, you examine the liquid with a microscope, you will not be able to trace any currents; hence the motion of liquids in diffusion is not in mass, but in molecules,—a kind of intermolecular motion. We learn that some liquids, even when stirred together, will not remain mixed; while others, whose densities are very different, when merely placed in contact with each other, slowly mix of themselves

§ 40. **Diffusion of Liquids through Porous Partitions — Osmose — Dialysis.** — *Very complex.* — **Experiment.** — Cut off the bottom of a conical-shaped bottle (or, better, use a glass funnel or lamp-chimney); fit to the neck of the bottle a cork having a glass tube passing through it (Fig. 17). Tie tightly over the bottom a piece of goldbeater's-skin or parchment paper. Fill the

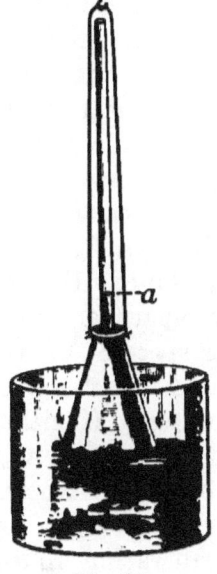

Fig. 17.

bottle with a concentrated solution of sulphate of copper, and press the cork into the bottle so that the liquid will stand a little way up the tube, say at *a*. Now suspend the apparatus in a vessel of water, so that the bottom may be covered. Leave it for an hour or more, and then carefully examine. What is the result? What does this experiment prove?

When liquids or gases force their way through porous septa,[1] and mix with each other, the diffusion is called *osmose*.[2] To distinguish the two opposite currents, the flow of the liquid or gas towards that which increases in

---

[1] Septum, *partition*.   [2] Osmose, *impulse*.

volume is called *endosmose*,[1] and the opposite current is called *exosmose*.[2]

It is found that crystallisable substances are the best subjects of osmose, while those which are usually amorphous, such as gelatine and gummy substances, are very little inclined to osmose. Those substances that pass readily through septa are called *crystalloids* ;[3] those that do not are called *colloids*.[4]

The principle of unequal diffusibility of liquids through septa finds important application in chemical and pharmaceutical laboratories. For example, from a rod (Fig. 18) is suspended a glass vessel having a bottom of parchment paper. Such a vessel is called a *dialyser*. In the dialyser is placed, for instance, the liquid contents of the stomach or intestines of a dead animal, suspected of containing some poison, and the vessel is floated in a vessel of water. If either arsenic or strychnine is present it will separate from the albuminous matter in the food, and pass through the septum into the water. The process of separating mixed liquids by osmose is called *dialysis*.

Fig. 18.

§ 41. **Free Diffusion of Gases**—*depends almost wholly on molecular motion.*—**Experiment.**[1]—Fill a test-tube with oxygen gas, and thrust in a lighted splinter ; the splinter burns much more rapidly than in the air. Fill another tube with hydrogen gas, and keep the tube inverted (for this gas, having only one-sixteenth the density of air, would very soon flow out of a tube held mouth upwards). Thrust in a lighted splinter ; the gas takes fire, and burns with a pale flame at the mouth of the tube. Next fill one tube with oxygen and the other with hydrogen gas, and place the mouth of the latter over the mouth of the former, as in Fig. 19. In about a

[1] Endosmose, *inward impulse*. [3] Crystalloid, *like crystal*.
[2] Exosmose, *outward impulse*. [4] Colloid, *like gum*.

minute apply a lighted splinter to the mouth of each tube (let the mouth of each tube be freely open to prevent accident). What is the result? What conclusion do you draw from it? Are oxygen and hydrogen nearly of the same density?

Experiment 2.—Fill a jar with carbon dioxide, a gas much denser than air, and uncover with the mouth upwards. How much of the gas remains in the jar at the end of ten minutes?

Many pairs of liquids do not diffuse into each other, but *every gas diffuses into every other gas*, and it is impossible to prevent two gases from mixing when placed in contact. (It is thought best to introduce the subject of diffusion of gases in this place, though it has little or no connection with the subject of adhesion. The explanation of diffusion must be deferred to its proper place in the chapter on Heat.)

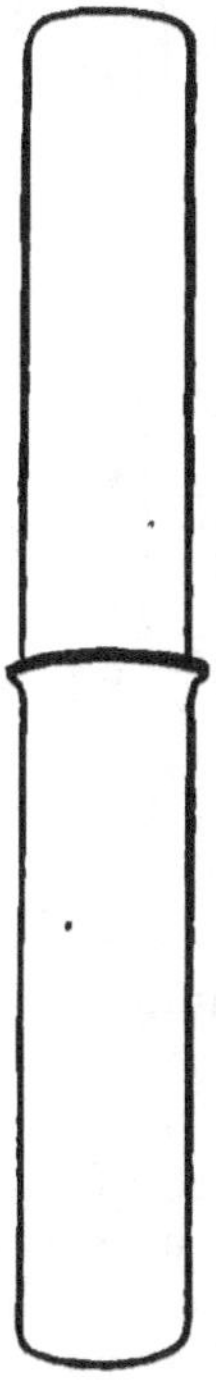

In consequence of this universal tendency to diffusion, gases will not remain separated,—*i.e.* a rarer resting upon a denser, as oil rests upon water. This is of immense importance in the economy of nature. The largest portion of our atmosphere consists of a mixture of oxygen and nitrogen gases. There are always present also small quantities of other gases, such as carbonic-acid gas, ammonia gas, and various other gases, which are generated by the decomposition of organic matter. These gases, obedient to gravity alone, would arrange themselves according to their density—carbonic-acid gas at the bottom, or next the earth, followed respectively by oxygen, nitrogen, water vapour, and ammonia. Neither animal nor vegetable life could exist in this state of things. But, in consequence of their diffusibility, they are found intimately mixed, and in the same relative proportions, whether in the valley or on the highest mountain peak.

Fig. 19.

§ 42. Diffusion of Gases through Porous Partitions—*probably depends on the number of molecules in a unit of volume, and the average speed of the molecules;* i.e.

upon the number of molecules striking upon a unit of area of the partition in a unit of time.

Experiment.—Take a thin, unglazed earthen cup, such as is used in Bunsen's battery, and plug up the open end with a cork through which extends a glass tube. Place the exposed end of the tube in a cup of coloured water. Lower a glass jar, filled with hydrogen or coal-gas, over the porous cup, as in Fig. 20. Observe carefully what takes place. How do you explain the result? Remove the glass jar. What happens? Does the result in this case lead to the same conclusion as that in the first?

An interesting modification of this apparatus is the *diffusion fountain* (Fig. 21). By passing the glass tube of the porous cup through the cork of a tightly-stopped vessel, and having another glass tube pass through another perforation in the same cork, water is forced out in a jet several feet in height, when the hydrogen jar is held over the porous cup.

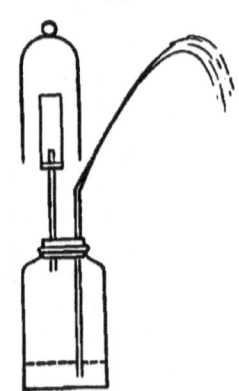

Fig. 20.

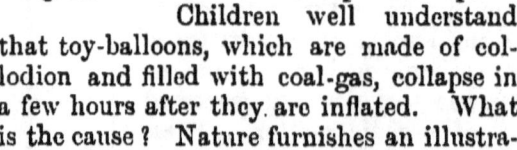

Fig. 21.

Children well understand that toy-balloons, which are made of collodion and filled with coal-gas, collapse in a few hours after they are inflated. What is the cause? Nature furnishes an illustration of osmose of gases in respiration. In the lungs the blood is separated from the air by the thin, membranous walls of the capillary veins. Carbonic-acid gas escapes from the blood through these septa, and oxygen gas enters the blood through the same septa.

CHAPTER II

KINEMATICS

§ 43. Motion and Rest Relative Terms.—To a person riding in a railway carriage, and confining his attention to objects in the carriage, everything appears to be at rest; but let him direct his attention to objects by the wayside, and at once he discovers that all objects in the carriage are in motion. Matter may be at rest with reference to certain objects, and in motion in regard to others. *Motion and rest are wholly relative terms*, and inapplicable to an object considered apart from all others. We cannot locate an object except with reference to another object, nor can we conceive of change or of permanence of position of an object, except in relation to some other object. The aeronaut, moving at the rate of sixty miles an hour, knows not that he is moving at all, till he looks away from his balloon, and sees cities and towns passing in panorama beneath him.

§ 44. All Matter is in Motion.—*There is no such thing as absolute rest in the universe.* There is no use for the word *rest*, except to indicate, with reference to each other, the condition of objects that are moving in the same direction and with the same speed. For example, a span of horses drawing a carriage are at rest with reference to each other and to the carriage. The phrase "at rest" can be used in only an extremely limited sense, and in common language refers only to the condition of an object with reference to that on which it stands, as a carriage, the deck of a ship, or the surface of the earth. It is only by putting entirely out of mind the motions of the earth itself that we can speak of any terrestrial object as being at rest.

§ 45. Speed—Uniform and Varied Speed.—All motion takes time; hence the term *speed*, which refers to *the space traversed in a unit of time.* Speed may be uniform or varied: *uniform*, when an object traverses successively equal spaces in all equal intervals of time; *varied*, when unequal spaces are traversed in any equal intervals of time. Varied speed may be quickened or retarded: *quickened*, when the spaces traversed increase at each successive interval of time; *retarded*, when they diminish. The speed of a train of cars in starting from a station is at first quickened, afterwards tolerably uniform, and, when the brakes are applied, it becomes retarded. Strictly speaking, all speeds are varied; there is no illustration of absolutely uniform speed in nature nor in art, though we may conceive of its possibility, and we have very closely approximated to it.

The *average speed* of a body during any interval is measured by the quotient arising from the division of the measure of the path traversed by the body in that interval by the measure of the interval. The unit of speed is therefore the speed at which a body traverses the unit of length in the unit of time. It will be seen that the unit of speed is made to depend upon the units of length and of time.

If a railway train travels half a mile in forty seconds, the measure of its average speed is $\frac{1}{2} \div \frac{40}{3600}$ or 45, if we choose a mile as our unit of length and an hour as our unit of time; but the measure of its speed is $\frac{5280}{2} \div 40$ or 66, if we choose a foot as our unit of length and a second as our unit of time. The speed is described in the one case as 45 miles per hour, in the other as 66 feet per second. It is, of course, the same speed in both cases, just as 4 shillings is the same sum of money as 48 pence.

If the speed of a point is *uniform*, its average speed during any interval is its *actual speed* (usually spoken of as the speed simply) at any instant of that interval. If, however, the speed of a point is not uniform, its actual speed at any instant is its average speed during an *infinitely short* interval which includes that instant. Thus, when we say that a railway train has a speed of 45 miles per hour, we do not necessarily imply that the train will run 45 miles during the next hour, or that it will continue running for an hour; we simply mean that during the short interval of our observation the average speed was 45 miles per hour, and that the train would run 45 miles were it to continue to run for an hour at the same speed it had at the moment referred to.

The change of speed during any interval is the difference between the initial speed and the final speed.

The average rate of change of speed during any interval in the quotient arising from the division of the change of speed during that interval by the measure of that interval. Thus if a speed changes from 15 feet per second to 60 feet per second during an interval of 15 seconds, the average rate of change is 3 feet-per-second per second. Thus the unit of rate of change of speed is a change of one unit of speed in one unit of time.

If the rate of change of speed is uniform the average rate of change of speed during any interval is also the actual rate of change of speed at any instant of that interval. If, however, the rate of change is not uniform, the actual rate of change of speed at a given instant is the average rate of change of speed during an infinitely short interval which includes that instant.

§ 46. Relations among Speed, Time of Motion, and Length of Path traversed.—Provided the speed is uniform, the foregoing relations are easily seen, and need not be discussed here. If, however, the speed is varied, these relations may be very complex. The only case of this kind that we shall consider is that in which the speed is increased by *equal* amounts during *equal* intervals or diminished by *equal* amounts during *equal* intervals, however short these intervals may be taken. In this case the speed is said to be *uniformly quickened*, positively or negatively.

A careful consideration of this case will enable the student to see that the *average speed* during any interval is half the sum of the initial and final speeds.

Let u denote measure of initial speed.
,, v ,, ,, final speed.
,, t ,, ,, time of motion.
,, k ,, ,, quickening during *one* unit of time.
,, s ,, ,, path traversed.

The following equations are obvious :—

$$v = u + kt \qquad (1),$$

$$s = \frac{u+v}{2} t \qquad (2).$$

The student acquainted with the elements of Algebra will observe that we have here two independent equations involving five symbols, and that provided three of these symbols be known, it is possible from these equations to find the other two. These equations will be found sufficient for the investigation of any case of uniformly quickened speed or of constant speed, since in the latter case $k = o$.

An example or two will serve to illustrate.

EXAMPLE 1.—A point starts from rest with a speed uniformly quickened at the rate of 10 feet-per-second per second, how far will it go in six seconds ?

The initial speed is 0.
,, final ,, 60 feet per second.
,, average ,, 30 ,,
therefore the space traversed in six seconds is 30 feet × 6 or 180 feet.

EXAMPLE 2.—A railway train, moving at the rate of 50 miles per hour, has the brakes put on, and its speed changes uniformly for one minute, when it is found to have a speed of 20 miles per hour. Find (a) its rate of change of speed, and (b) the distance traversed in the minute.

Taking one hour as our unit of time, one mile as our unit of space, and therefore a mile an hour as our unit of speed, we have given—

$$u = 50,$$
$$v = 20,$$
$$t = \frac{1}{60}.$$

Substituting these values in our fundamental equations, we have—

$$20 = 50 + \frac{k}{60} \qquad (1),$$

$$s = \frac{50 + 20}{2} \times \frac{1}{60} \qquad (2).$$

From (1), $1200 = 3000 + k$,
$$\therefore k = -1800.$$

From (2), $120\ s = 70$,
$$\therefore s = \frac{70}{120},$$
$$= \frac{7}{12}.$$

Hence we see that the speed of the train was *diminished* at the rate of 1800 miles-per-hour per hour, and that the train ran $\frac{7}{12}$ of a mile during the minute.

Let us solve the same problem, using a foot as the unit of space, a second as the unit of time, and, therefore, a foot-per-second as our unit of speed.

In this case $u = \dfrac{5280 \times 50}{60 \times 60} = \dfrac{220}{3}$,

$v = \dfrac{5280 \times 20}{60 \times 60} = \dfrac{88}{3}$,

$t = 60$.

Substituting these values in our fundamental equations, we have—

$$\dfrac{88}{3} = \dfrac{220}{3} + 60k \qquad (1),$$

$$s = \dfrac{\dfrac{220}{3} + \dfrac{88}{3}}{2} \times 60 \qquad (2).$$

From (1), $88 = 220 + 180k$,

$$\therefore k = \dfrac{88 - 220}{180} = -\dfrac{11}{15}.$$

From (2), $s = \dfrac{220 + 88}{6} \times 60$,

$\qquad\qquad = 3080$.

Hence we see that the speed of the train was *diminished* at the rate of $\tfrac{11}{15}$ of one foot-per-second per second, and that the train ran 3080 feet during the minute.

The student should thoroughly satisfy himself that these answers, though different in *form*, are virtually the same as the answers obtained by the first solution. He will probably find some difficulty in reconciling the two statements regarding the rate at which the speed of the train is diminished, but he should not pass this point until he is satisfied.

NOTE.—The recommendation of Professor Tait of Edinburgh has been followed in the use of the word "speed" to denote the length of the path traversed in a unit of time without regard to direction of motion, while the word "velocity" is reserved for the expression of that which involves the ideas of speed and of direction of motion conjointly.

§ 47. **Velocity.**—If at the beginning of an interval a moving point is at A and at the end of the interval it is at B, the *average velocity* of the point during this interval is a quantity whose magnitude is measured by the quotient arising from the division of the measure of the line AB by the measure of the interval, and whose direction is that of the line AB.

If the velocity of a moving point is uniform (that is, if it is moving with uniform speed and in a straight line) its average velocity during any interval is its *actual velocity* (usually spoken of as the velocity simply) at any instant of that interval. If, however, the velocity of a moving point is not uniform, its actual velocity at any instant is its average velocity during an *infinitely short* interval which includes that instant.

Since a velocity possesses the elements of magnitude and of direction, it is conveniently represented by a straight line, the length of the line representing the magnitude of the velocity and the direction of the line representing the direction of the velocity.

§ 48. Composition and Resolution of Velocities.

—To compound two velocities (that is, to find a single velocity equivalent to the combined effect of two velocities) we have the following obvious construction. Let one of the given velocities be represented in magnitude and direction by AB, and let the other given velocity be represented, on the same scale, in magnitude and direction by BC. Then AC represents (still on the same scale) in magnitude and direction the resultant velocity.

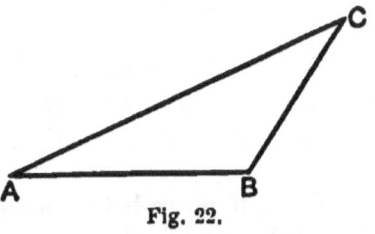

Fig. 22.

To understand this, imagine a steamer to be sailing with a velocity represented by AB, and at the same time imagine a man to be walking on its deck with a velocity represented by BC. A may be taken to represent the position of the man at the beginning of his walk. At the end of the man's walk the point A is at B in consequence of the velocity of the steamer with respect to the earth, and the man is at C in consequence of his simultaneous velocities with respect to the earth, resulting from the motion of the steamer through the water and from his own walk along the steamer's deck. Thus AC represents the *resultant velocity* of the man with respect to the earth. The student must observe that when it is said that a velocity is represented by a line it is to be understood that this line represents the velocity both in magnitude and direction.

The foregoing proposition is called the triangle of velocities. Another obvious way of stating it is indicated by completing the parallelogram of which AC is a diagonal (Fig. 23). *Let two velocities to be compounded be represented, on the same scale, by two lines drawn from the same point. Complete the parallelogram of which these lines are adjacent sides, and that diagonal of this parallelogram which passes through this point shall represent (still on the same scale) the resultant velocity.* This proposition is called the parallelogram of velocities, and it should be thoroughly understood.

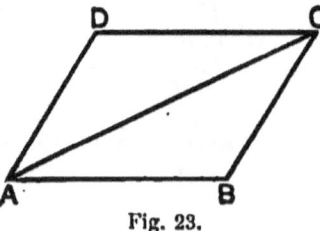

Fig. 23.

The difference between two given velocities is evidently that velocity which, compounded with one of the given velocities, will produce as a resultant the other given velocity. Hence the change which a velocity undergoes during any interval when its magnitude and direction at the beginning, and also at the end, are known, is readily obtained by the following construction.

Draw AC to represent the velocity at the beginning of the interval. Draw AB to represent, on the same scale the velocity at the end of the interval. Then CB represents (still on the same scale) the change which the velocity has undergone during the interval.

By the triangle of velocities, the velocities AC and CB have as a resultant the velocity AB. Hence CB represents the change that must be made in the velocity AC to produce the velocity AB.

It will be observed that the change of velocity is different from the change of speed. In this case the change of speed is represented by the difference between AC and AB.

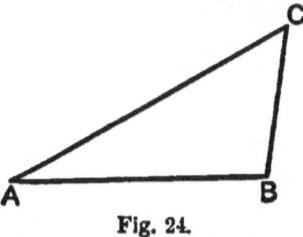
Fig. 24.

§ 49. Rate of Change of Velocity.—Rate of change of velocity is called *acceleration*. The *average* acceleration during any interval is a quantity whose magnitude is the quotient arising from the division of the change of velocity undergone by the moving point during that interval by the measure of that interval, and whose direction is the direction of that change of velocity. Thus (Fig. 24) if the velocity of the moving point changes from AC to AB in t units of time, $\dfrac{CB}{t}$ is the magnitude of the average acceleration during this interval and CB is the direction of this acceleration.

Generally the average acceleration of a moving point varies, both in magnitude and in direction, with the interval to which it applies. In the particular case in which it varies neither in magnitude nor in direction the point is said to move with a *uniform acceleration*.

The *actual acceleration* (commonly spoken of as acceleration simply) at any given instant is the average acceleration during an infinitely short interval which includes that instant.

The specification of an acceleration involves specification both of its magnitude and of its direction. Its direction may be described as the direction of a straight line is described. Its magnitude is evidently a quantity of the same kind as a rate of change of speed (§ 45), and may therefore be described in terms of the unit of rate of change of speed.

QUESTIONS AND PROBLEMS

1. A railway train travels over 150 miles in 5 hours, 40 minutes; what is its average speed, a foot and a second being the units of space and of time?

2. A point passes uniformly over 100 yards in 30 minutes; what is the numerical representation of its speed; according to the usual conventions respecting the units of space and of time?

3. A man walks with a speed represented by 3, and he finds that he walks 5 miles in 2 hours; if 1 foot be the unit of space, what is the unit of time?

4. A body whose speed is represented by 3, moving uniformly, describes 64 feet in 10 seconds; the unit of space is 1 foot; what is the unit of time? By what would its speed be represented, if the units of space and of time were 3 yards and 2 minutes?

5. A point is observed to describe uniformly a feet in n seconds; supposing the unit of time to be 1 minute, what must be the unit of space in order that the numerical value of the point's speed may be 1?

6. A body describes 72 feet, while its speed increases from 16 to 20 feet per second; find the whole time of motion and the rate of change of speed.

7. The speed of a particle moving with uniformly increased speed on passing a given point P is 10 feet per second; at a distance of 15 feet from P it is 20 feet per second; find its rate of change of speed, also its distance from P when at rest.

8. A particle is proceeding with uniform velocity along a certain straight line, and another velocity is communicated to it such that it moves along a line making 60° with its former direction and with unaltered speed. Determine the communicated velocity.

9. If a particle has a velocity of 20 where a day and a mile are the units of time and of space, and a velocity perpendicular to the former of $2\frac{4}{15}$ when the units are a minute and a foot,

find the resultant velocity when an hour and a yard are the units.

10. A boat is rowed across a river (2 miles wide) in a direction making an angle of 60° with the bank and against the stream. The boat travels, with reference to the water, at the rate of 5 miles per hour, and the river flows at the rate of three miles per hour. Find at what point of the opposite bank the boat will land.

·11. In the preceding problem, at what angle to the bank should the boat be rowed in order to land at the opposite point of the opposite bank?

12. A point is moving in the circumference of a horizontal circle and with constant speed b. Find its total change of velocity (both in magnitude and in direction) in the interval between the moment at which it is moving due north and the moment at which it is moving due east. Solve this problem, assuming that the point at the beginning of the interval is (a) at the west side of the circle, (b) at the east side.

13. If velocities be separately communicated to a particle on the circumference of a circle, such that they would respectively cause the particle to move to the circumference along chords at right angles to each other in the same time, show that when the velocities are communicated to it simultaneously, the time of its reaching the circumference will be unaltered.

CHAPTER III

DYNAMICS

In the chapter on kinematics we have considered motion pure and simple, without regard to that which was moving or to the cause of the motion. That is, without regard to matter or to force. In this chapter we shall investigate the relations existing among matter, force, and motion.

§ 50. Inertia.—What is it that sets in motion that which was previously at rest, or in any other way changes the motion of a portion of matter? We may call it *force*; but what idea does this term convey? Let us question our own experience. We leave an apple lying upon a table; have we not entire confidence that it will continue to lie there, unless disturbed by some *other* body? If on returning we find it gone, are we not sure that it has been removed by the action of some body other than itself? An apple falls to the ground, and although the action is one of the most mysterious in all nature, yet do we not almost instinctively trace the cause to some action between the apple and the earth? The ball at rest is put in motion by a bat; but must not the bat first be put in motion? And when we find the cause of its motion, is it not an antecedent motion in some other object? We conclude, then (1), that *motion cannot* *originate in an object isolated from all others, but it always arises from* MUTUAL *action between at least two bodies.*

Again, the bat, having received motion, is capable of imparting motion to the ball; but, having set in motion one ball, is it equally capable of putting in motion another ball? Can a mass impart motion and retain all its own motion? Or is it like a commercial transaction, a trade, to which there are two parties, one a buyer and the other a seller? that is, of the nature of a transfer, which should be entered on the debit side of one's account, and the credit side of the other's? From a careful consideration of these questions we conclude (2) that *change of motion in one body is caused only by another body's parting with some of its power of producing change of motion.*

If a sledge, on which a child is sitting, is *suddenly* put in motion, what is the result? If the child and sledge are both in motion, and the sledge is *suddenly* stopped, what happens? How is it in the first case if the sledge is started *slowly*? How in the second case if the sledge is *gradually* stopped? We see (3) that *masses of matter receive motion gradually and surrender it gradually.*

Even very small bodies require time to start and to stop. The sand-blast, employed for engraving figures on glass, furnishes a fine illustration of this fact. A box of fine quartz-sand is placed in an elevated position. A long tube extends vertically down from the bottom of this box. The plate of glass to be engraved is covered with a thin layer of melted wax. When cool, the design is sketched with a sharp-pointed instrument, in the wax, leaving the glass exposed only where the lines are traced. The plate is then placed beneath the orifice of the tube, and exposed to a shower of sand. The speed of the sand-grains is not at its maximum at the start, but is constantly increased till they reach the plate, where their speed in turn is gradually checked. The wax, on account of its yielding nature, gradually brings them to rest without being chipped; but the glass, notwithstanding its hardness, cannot stop them quite at its surface; and, therefore, being very brittle, it suffers a chipping action from the sand. Thus the soft wax affords a protection from the action of the falling

F

sand to all parts except those intended to be cut. A still greater speed is generally given to the sand by steam blown through the tube. For this reason the apparatus is called a *sand-blast*. Hard metals like steel are engraved in the same manner. Yet the hand may be held in the blast several seconds without injury. (What is the difference in the effects of catching a cricket-ball with hands held rigidly extended, and with the hands allowed to yield somewhat to the motion of the ball?)

Roll a marble on a carpet, what happens? roll it on a smooth marble floor, what is the result? On a perfectly smooth surface how long would it roll? If we could provide such a surface, and dispense with the resistance of the air, how long would it roll? These conditions are impracticable? True. But have not the heavenly bodies rolled for millions of years through frictionless space, unchecked because unimpeded?

Motion unobstructed is *perpetual*. Motion undisturbed is in a *straight line*. Along which will a marble roll more nearly in a straight line, along a smooth or along a rough floor? What if the floor were perfectly smooth?

The foregoing observations may be summed up by saying that *matter possesses a property by virtue of which it maintains* CONSTANT *its condition of motion, unless influenced by some* EXTERNAL *cause*. This property of matter has been called INERTIA. The statement of the fact that matter possesses this property has been called *the law of inertia*.

This fact is also called the first law of motion, and is frequently enunciated thus:—

§ 51. First Law of Motion.—*A body at rest remains at rest, and a body in motion moves with uniform velocity, (that is, with uniform speed and in a straight line) unless acted upon by some external force to change its condition.*

This law is briefly summarised in the familiar expression, "perpetual motion." "Is perpetual motion possible?" has been often asked. The answer is simple,—

Yes, more than possible, *necessary, if no force interferes to prevent*. What has a person to do who would establish perpetual motion? Isolate a moving body from interference of *all* external forces, such as gravity, friction, and resistance of the air. *Can the condition be fulfilled?*

§ 52. Relation between a Force and the Change of Motion it produces.—Having learned, from experience, that a body undergoes no change of motion except from the action of an external force, we, naturally, next proceed to investigate the relation between any force and the change of motion it produces. To make our work as simple as possible let us begin by examining the motion of a body under the influence of *only one* external force. It is further desirable that this force shall be *constant*—that is, of unchanging intensity and direction. Now the *weight* of the same body at different heights (within moderate limits) at the same point on the earth's surface is practically constant, as may be shown by suspending the body by means of a coiled spring, and observing the extent to which its weight extends the spring when the body is thus suspended at different heights. Again, by suspending the body (at different heights) from the end of a long string, we find that the *direction* of its weight at different heights above the same point on the earth's surface is the same. Hence, if we can watch a body moving near the surface of the earth under the influence of its weight alone, we shall be able to learn the effect of a force constant both in intensity and in direction. Can we secure this condition? Not without removing the air from the space in which the body is to move, for, as is easily seen, as soon as a body begins to move through the air, the air exerts a force upon it in a direction opposite to that in which the body is moving, and the intensity of the force increases with the speed of the body.

Since we cannot readily remove the air, let us follow

the advice given in § 1, and so arrange our experiment that the weight of the body shall be very great in comparison with the force which the air exerts upon the body as it moves through it. In this case we shall not introduce an important error by neglecting to consider the force exerted by the air, and assuming that the whole effect observed is due to the weight of the body. We shall best secure the condition desired by choosing a dense body, say a lump of lead, and by confining our observations of its motion to a period in which its speed is at no time very great. Why should we choose a dense body? Why should we not observe its motion through the air at very great speed?

Since it is *change of motion* that we wish to observe, let us still further simplify our experimentation by beginning with a body initially at rest. (Why?)

Experiment 1.—Secure two pieces of lead weighing 1 oz. and 2 oz. respectively. Drop the first from the height of a few feet. What do you observe regarding the *direction* of its motion? What do you observe regarding its *speed*? Does the direction of its motion change? What relation does it bear to the direction of the weight of the body? Does the speed of the body change? What is the nature of this change so far as you can observe? Embody the result of your observations in a statement regarding the *change of velocity* (§§ 48, 49) which the body undergoes.

Drop the two pieces, both at the same instant and from the same height. Does one reach the ground before the other? Is there any difference between the *change of velocity* that the one undergoes and the *change of velocity* which the other undergoes? Does the same force produce this change of velocity in the one case as in the other? What relation do these two forces bear to each other? In what respect does the one body differ from the other body? The bodies are of the same material; do they contain equal quantities of matter? What is their relation in this respect?

From these observations some important truths may be learned.

(1) *A constant force acting upon a body produces a change of velocity in the* DIRECTION *in which the force acts.*

(2) *This change of velocity varies with the* TIME *during which the force is acting.*

(3) *If the* MASS *of a body is increased the force necessary to produce (in a given time) a given change of velocity must be increased at the same rate.*

Our knowledge of the second of these truths is, as yet, indefinite, since we have made no exact measurements of the speed of the falling bodies at different points. Our experiment has been *qualitative*, not *quantitative*.

Without apparatus difficult to secure we cannot accurately measure the speed of a body falling freely, even through a short distance. The speed soon becomes too great. Arrange a contrivance like the following:—

Experiment 2.—Take a smooth straight board (Fig. 25), about 4^m long, and place it so that one end shall be about 4^{cm} higher than the other. Suspend within easy view a string (about 1^m long) and ball, as a pendulum, and draw a line

Fig. 25.

just under the ball. Set it in vibration, and, at the instant the ball crosses the line, let a marble begin to roll down the inclined plane. Let another person mark the point on the board that the ball reaches at the end of one swing of the pendulum—that is, at the instant the ball crosses the line next time. Repeat the operation several times, and mark the point that it reaches at the end of the second, of the third, and of the fourth swings. Verify your results by several trials. If you find any difference, take the average of the distances traversed by the marble in each period. Take the period of one swing of your pendulum as the unit of time, and the space traversed during the first swing as the unit of space.

Arrange the results of your observations in a tabulated

form. If the experiment is conducted with care you will obtain the following results :—

No. of units of time.	Total distance passed over.	Distance passed over in last unit. *Average velocity during last unit.*	Increase of velocity in each unit, i.e. *acceleration.*
1	1	1	2
2	4	3	2
3	9	5	2
4	16	7	8
etc.	etc.	etc.	etc.

In the foregoing experiment, why must the board be smooth? Why must it be straight? What relation does the experiment prove to exist between the *change of velocity* produced by a constant force and the *time* during which the force is acting to produce this change? Now state the second of the three truths learned from our first experiment in a *definite* form. Does it appear from Experiment 2 that the change of velocity produced by a given force in a unit of time depends upon whether the body acted upon is initially at rest or initially in motion? What is the change of velocity the marble undergoes during the first unit? What during the second? What during the third?

What we have learned from our two experiments may be summed up as follows :—*The effect of a constant force acting alone is the same whether the body acted upon is initially at rest or initially in motion; and that effect is to produce in the body acted upon a change of velocity whose direction is the direction of the force, and whose magnitude varies directly as the intensity of the force, directly as the time during which the force acts, and inversely as the mass of the body acted upon.*

We may, of course, make use of what we have already learned in proceeding with any further investigations.

Let us try to find an answer to the question—*Is the*

effect of a force the same whether acting alone or acting in conjunction with other forces?

Experiment 3.—Obtain from an ironmonger two spring-balances. From pegs A and B (Fig. 26), in the frame of a blackboard, suspend a known weight W (say) 10 lbs., by means of two strings connected at C. In each of these strings insert a spring-balance, as in the figure. Allow the combination to come to rest. Note carefully the indication of each spring-balance. Mark the point C on the blackboard and take down the apparatus.

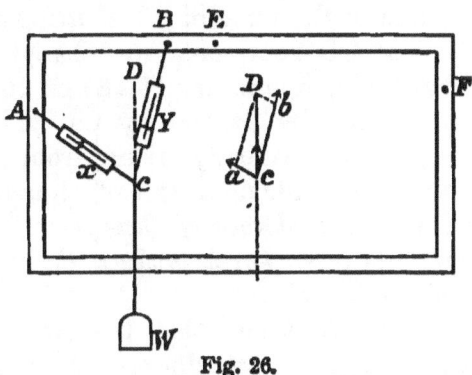

Fig. 26.

Join CA and CB and draw a vertical line through C. From CA cut off Ca containing as many inches as the spring-balance placed in CA indicated pounds, and from CB cut off Cb containing as many inches as the spring-balance placed in CB indicated pounds. Complete the parallelogram of which Ca and Cb are adjacent sides and draw the diagonal CD. Carefully determine the length and the direction of CD. Repeat this experiment many times, using different weights, until you are able from your observations to draw, with confidence, a *general conclusion*. Enunciate this general conclusion.

Now since, in our experiment, the particle at C undergoes no change of motion, it is obvious that the *effect* of the force W (10 lbs.) must be exactly equal and opposite to the *combined effects* of the pulls of the spring-balances (say x lbs. and y lbs. respectively). We have already learned what the effect of a force acting alone is; and we have already learned (§ 48) what the resultant of two velocities is. The change of velocity produced by a constant force acting alone, during a given time on a given mass, is in the direction of the force, and varies directly as the intensity of the force. Hence Ca properly represents the change of velocity arising from the action on a unit mass during a unit of time of the force x lbs. *acting alone.* Likewise Cb represents the change of velocity arising from the action on a unit of mass during a unit of time of the force of y lbs. *acting alone.*

Also CD represents the change of velocity resulting from the compounding (§ 48) of the changes of velocity represented by Ca and by Cb. Therefore CD represents the change of velocity arising from the combined action on a unit mass during a unit of time of the forces x lbs. and y lbs., *provided each of these forces, when acting in conjunction with other forces, produces the same effect as it produces when acting alone.*

Now we know that the two forces x lbs. and y lbs. acting together *actually produce* an effect equal in magnitude and opposite in direction to that produced by 10 lbs. acting vertically downwards. Does the line CD represent a change of velocity equal and opposite to that produced by the action on a unit of mass during a unit of time of a force of 10 lbs. whose direction is vertically downwards? In other words, in our experiment, is the line CD vertical in direction and 10 inches in length?

Having satisfied ourselves on this point, we are now prepared to make a full statement regarding the effect of a constant force. What we have learned may be conveniently enunciated thus:—

§ 53. Second Law of Motion.—*The effect of a uniform force is the same whether the body acted upon is initially at rest or initially in motion, whether the force acts alone or in conjunction with other forces; and this effect is to produce in the body acted upon a change of velocity, whose direction is the direction of the force, and whose magnitude varies directly as the* INTENSITY OF THE FORCE, *directly as the* TIME *during which the force continuously acts, and inversely as the* MASS *of the body acted upon.*

QUESTIONS

Of what truths stated in the "Second Law of Motion" are the following facts examples?

1. Two iron balls of different sizes dropped from the same height at the same moment reach the ground together.
2. A ball let fall in a railway carriage running 40 miles an hour falls to the floor with the same rapidity as if let fall inside a house.
3. The speed of a body falling freely under the action of its weight has been found to increase 32·2 feet-per-second per second.
4. If a guinea and a feather are enclosed in a long glass tube, the air pumped out and the tube closed, it is observed on turning the tube, so as to allow them to fall from one end to the other, that the guinea and the feather fall at the same rate. Why do they not so fall through the air?
5. If two forces of 10 lbs. each act upon a particle at right angles to each other, it is found that a force of $10\sqrt{2}$ lbs. is required to counterbalance them, and that it must act along a line which (produced backwards) bisects the angle between the two equal forces.

§ 54. Falling Bodies.—Having learned from observations what the effect of a force is, we are now in a position to make use of this knowledge to assist us in some interesting calculations. Let us first consider a body, initially at rest, falling freely near the surface of the earth. In this case the force acting upon the body is constant both in magnitude and in direction, and therefore the body undergoes a *uniform acceleration* (§ 49). Since the body is initially at rest its acceleration is in the direction of its motion at every point, and therefore its *speed* changes at a uniform rate. We can therefore investigate its motion after the manner indicated in § 46. In like manner we may treat motion from rest under the sole influence of any constant force.

The acceleration due to gravity is found to differ slightly at different points on the earth's surface (§ 21). In England it is 32·2 feet-per-second per second. It is customary to represent this by g. If we have to deal with a body under the simultaneous influence of two or more forces, we must bear in mind the fact that each produces the same effect as if it acted alone.

PROBLEMS

(Solve these problems from 1 to 12 in both the metric and the English measures.)

1. Disregarding the resistance of the air, what distance will a body fall from a state of rest in five seconds?
2. What distance will it fall during the fifth second?
3. What is its velocity at the end of the fifth second?
4. A stone, dropped from a balloon, strikes the ground in seven seconds. How high is the balloon?
5. Under the influence of a constant force a body moves from rest 500^m in a minute. How far will it go in an hour?
6. What will be its velocity at the end of the first half-hour?
7. How far will it move during the fifty-ninth minute?
8. A body falls from rest to the ground in four seconds; meantime it is acted on by a constant force which causes it to move in a horizontal direction 2^m in the first second. Where will it strike the ground?
9. What is its horizontal velocity at the end of the fourth second?
10. What is its vertical velocity at the end of the fourth second?
11. With what vertical velocity must a body start that it may continue to ascend during three seconds?
12. How far does it rise during the first second?
13. A stone is thrown upwards with a velocity of 161 feet-per-second, and two seconds afterwards another stone with a velocity of 225·4 feet-per-second; when and where will the stones meet?
14. In Question 13, if the stones had been projected downwards, when and where would they have met?
15. A stone is thrown upwards with a velocity of 100 feet-per-second, and three seconds afterwards another stone is thrown upwards with a velocity of 200 feet-per-second; find their distance apart, four seconds after the first stone started.
16. A stone is thrown downwards from the top of a tower

three seconds after one is dropped; with what velocity must it be thrown so as to overtake the other in four seconds?

17. How high will a body rise which is thrown vertically upwards, and returns to the place of starting in ten seconds?

18. A stone falling for two seconds breaks a pane of glass, and thereby loses one-half of its velocity; find the space described in six seconds from starting.

19. A body is projected upwards with a velocity of 200 feet-per-second; when and at what height will its velocity be 80 feet-per-second?

20. A stone falling from rest for four seconds passes through a pane of glass, thereby losing one-third of its velocity, and reaches the ground three seconds afterwards; find the height of the glass.

21. Gravity at the surface of the planet Jupiter being about 2·6 times as great as at the surface of the earth, find the distance traversed and velocity acquired by a body falling from rest for five seconds towards Jupiter from a point near its surface.

22. A stone is projected upwards with an initial velocity of 60 feet-per-second, and has a negative acceleration of 10 feet-per-second per second; how high will it rise?

§ 55. Projectiles.—Experiment.—Take a bottomless tin can A (Fig. 27), and connect an india-rubber tube C, 2^m long, with a glass tube passing through a stopper at B, and insert a short glass tube at D. Keep the can filled with water, bend the lower part of the india-rubber tube at D, so as to direct the stream at different angles of elevation, and observe the peculiarities of the curves formed by the streams, and the different vertical and horizontal distances reached by each.

In this experiment you have a miniature representation of the paths of all projectiles,[1] such as cannon-balls, stones thrown from the hand, etc. The horizontal distance that the projectile attains is called its *range*. Theoretically, the greatest range is obtained at an angle of 45°; but practically, on account of the resistance of the air, it is at a little less than 40°.

Every projectile is acted upon by two forces: (1) the

---

[1] Projectile, *a body thrown*.

force of gravity, and (2) the resistance of the air. It also has a certain speed and direction at the instant of projection. If this speed and direction are known, and the

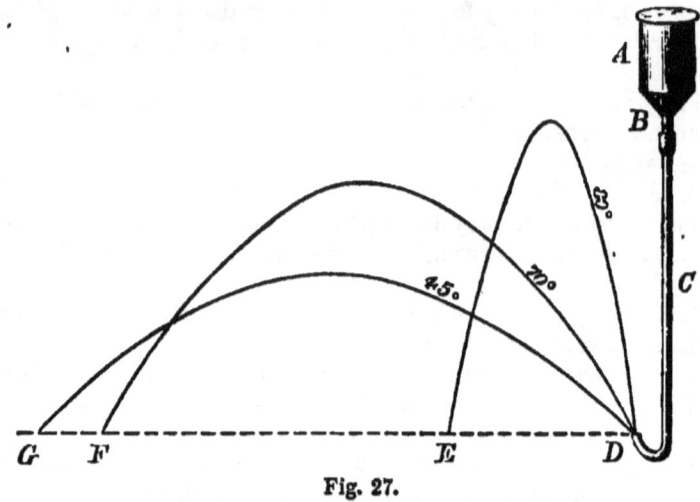

Fig. 27.

resistance of the air is disregarded, the path of a projectile can be determined. Thus, suppose that a projectile is thrown from A (Fig. 28) at an angle of 45°, that it is in

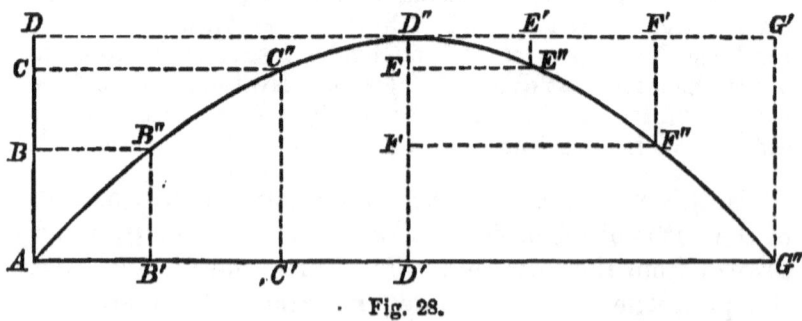

Fig. 28.

the air six units of time, and that the vertical heights reached at the end of the first three units successively are B, C, and D. Its horizontal motion, if unimpeded, is uniform, and the corresponding points reached in that direction at the same moments are (say) B', C', and D'. Combining these two motions, we obtain the points B", C", and D", reached by the projectile successively, at the end of each of the first three units of time. The force of gravity constantly acting to change its direction, it must describe, during the first three units, the curved line AB"C"D". Since the time of ascent and descent are equal, it must reach its greatest vertical height at the end of the third unit, when it begins its descent. The path of descent D"E"F"G" is found in a similar manner The path thus described is known as a *parabolic curve*. The curve D"E"F"G" represents also the path of a projectile thrown from D", in the direction of the line D"G', with a horizontal velocity that would cause it to reach G' at the end of the third unit of time.

QUESTIONS

1. Why are AB', B'C', and C'D' (in the figure) equal?
2. What are the relative lengths of DC, CB, and BA?
3. If the initial speed of this projectile is $96 \cdot 6 \sqrt{2}$ feet-per-second, what is its speed at the point D"?
4. How high does it rise?
5. What is its speed at the end of the first second?
6. What is its range?

§ 56. Component Velocities.—It is obvious that a velocity produces no displacement in a direction at right angles to itself. Hence to find the *effective component* in a given direction of a given velocity we have the following construction. Draw AB (Fig. 29) to represent, on some convenient scale, the given velocity in magnitude and direction. Let AC be the direction in which we wish to find the effective component of this given velocity.

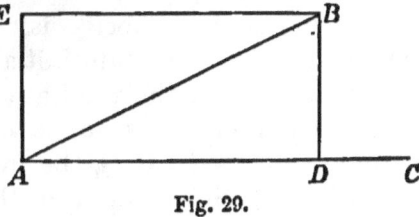

Fig. 29.

From B draw BD at right angles to AC and complete the parallelogram ADBE. It has been shown (§ 48) that AB represents the resultant of the velocities represented by AD and by AE. Now since AD and AE are at right angles to each other, a movement along one causes no displacement in the direction of the other. Therefore AD represents the effective component of the given velocity in the direction AC.

The ratio of AD to AB depends, of course, upon the angle BAD. If the angle BAD is 45°, AD = BD (Euclid I. 32, 6) and therefore $\frac{AD}{AB} = \frac{1}{\sqrt{2}}$ (Euclid I. 47). If the angle BAD is 30°, the triangle ABD is half of an equilateral triangle (Euclid I. 5, 32) and BD = ½ AB. Therefore in this case $AD = \frac{\sqrt{3}}{2}$ AB.

§ 57. Investigation of Projectiles continued.— In investigating the paths of projectiles we shall find it convenient to consider the horizontal and vertical motions separately.

If we neglect the resistance of the atmosphere, which, for the sake of simplicity, we shall do, the horizontal motion is a uniform motion; while the vertical motion is, during the upward flight, a uniformly retarded, and, during the downward flight, a uniformly accelerated motion. Why?

The horizontal velocity is, of course, the horizontal component of the absolute initial velocity, and the initial vertical velocity is the vertical component of the same.

The chief elements to be considered are the initial absolute velocity (that is, the initial velocity in the line of projection), the direction of the projection, the time of flight, the highest point of path, the range, the absolute velocity at any given moment after projection, and the position of the projectile at any given moment after projection.

Example.—Let a cannon-ball be fired on a horizontal plane, with a velocity of 1544 feet-per-second, at an angle of 30° above the horizon.

The horizontal velocity $= 772\sqrt{3}$ feet-per-second.

,, initial vertical ,, $= 772$,, ,, (56).

The acceleration due to gravity $= 32\frac{1}{6}$ feet-per-second per second (§ 54) vertically downward.

The ball will continue to rise until the action of gravity reduces its vertical velocity to zero, and it will then continue to descend until it strikes the plane. Since its vertical motion is influenced by gravity during the fall as well as during the rise, the time of falling is the same as the time of rising.

Hence, the time of flight $= \dfrac{772}{32\frac{1}{6}} \times 2 = 48$ seconds.

Since during all this time the horizontal velocity is uniform, and $772\sqrt{3}$ feet-per-second, therefore the range $= (772\sqrt{3} \times 48)$ feet.

To find the highest point of path, find the height to which a body having an initial upward vertical velocity of 772 feet-per-second will rise in 24 seconds.

Therefore, highest point of path $= \left\{ 772 \times 24 - \dfrac{32\frac{1}{6}}{2} \times (24)^2 \right\}$ ft.

$= 9264$ feet.

To find the position of the ball at the end of 14 seconds, find its height at this moment by considering its vertical motion, and find its horizontal position at the same moment by considering its horizontal motion. Having its height and horizontal position, its absolute position is of course fixed. Work this out.

To find the absolute velocity of the ball at the end of 14 seconds, square the measures of its vertical and horizontal velocities at this moment, and extract the square root of the sum of these squares. Why?

The horizontal velocity is uniformly $772\sqrt{3}$ feet-per-second.

The vertical velocity at end of 14 seconds is $(772 - 32\frac{1}{6} \times 14)$ feet-per-second $= 321\frac{2}{3}$ feet-per-second. Hence the absolute velocity at the end of 14 seconds $= \sqrt{(772\sqrt{3})^2 \times (321\frac{2}{3})^2}$ feet-per-second.

The student who keeps in mind the fact that the vertical and horizontal velocities of a projectile may be considered independently, will not find difficulty in dealing with any similar problem.

QUESTIONS

1. A particle is projected at an angle of 45° with the horizon from a point on a horizontal plane, with a speed 1000 feet-per-second. Find its range, and find its distance from the point of projection at the end of five seconds.

2. A particle is projected at an angle of 30° from a point on a horizontal plane, and its total range is 5000 feet. What is the speed of projection?

3. A particle is projected at an angle of 30° from a point on a horizontal plane, and the highest point it reaches is 500 feet above the plane. What is the speed of projection, and the total range?

4. A ball is fired from the ground at an angle of 45°, so as just to pass over a wall 10 feet high at a distance of 260 feet. Where will it strike the ground?

5. "Swift of foot was Hiawatha;
He could shoot an arrow from him,
And run forward with such fleetness
That the arrow fell behind him!
Strong of arm was Hiawatha;
He could shoot ten arrows upward,
Shoot them with such strength and swiftness,
That the tenth had left the bowstring
Ere the first to earth had fallen."

Neglecting the resistance of the air, taking $g = 32$, supposing one second to elapse between the discharge of each of the ten arrows, and making Hiawatha shoot at his longest range, find the speed he must have been capable of exceeding.

§ 58. Curvilinear Motion.—If a body is moving in a curve under the influence of a single external force, what is necessary regarding the direction of this force? If the body is moving with constant speed, what is necessary regarding the direction of this force? In the latter case can the force have any component in the direction in which the body is moving?

Suppose a ball at A (Fig. 30), suspended by a string from a point d, to be struck by a bat, in a manner that would cause it to move in the direction Ao. At the same time it is restrained from taking that path by the tension of the string, which operates like a force drawing it toward *d*. It therefore takes, in obedience to the two forces, an intermediate course toward *c*. At *c* its motion is in the direction *cn*, in which path it would move, but for the string, in accordance with the first law of motion. Here, again, it is compelled to take an intermediate path toward *e*. Thus, at every point, the tendency of the moving body is to preserve the direction it has at that point, and consequently to move in a straight line. The only reason it does not so move, is that it is at every point forced from its natural path by the pull of the string. But if, when the ball reaches the point *i*, the string is cut, the ball, having no force operating to change its motion, continues in the direction in which it is moving at that point; *i.e.* in the direction *ih*, which is a tangent to its former circular path.

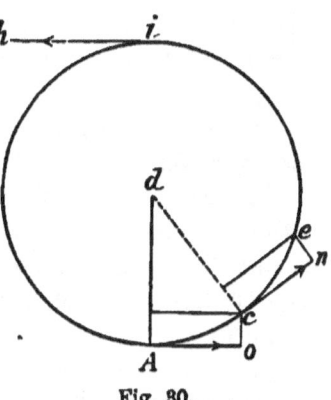

Fig. 30.

This tendency of a body moving in a curvilinear path to fly off in a straight line has been erroneously attributed to a supposed "centrifugal force," which is constantly urging it away from the centre, its escape being prevented only by a force pulling it toward the centre.

Centrifugal force has in reality no existence; the results that are commonly attributed to it are due entirely to the tendency of moving bodies to move in straight lines in consequence of their inertia. If a moving body is to describe a curvilinear path, a force called a *centripetal force* must be constantly applied to it at an angle to its otherwise straight path.

The greater the speed of the moving body, the greater must be the force applied to produce a given departure from a straight line. This may be shown by suspending a weight to a dynamometer, and swinging them about the hand. If, when thirty revolutions are made in a minute,

G

the force, as indicated by the dynamometer, is 4 pounds, then, on doubling its speed, the force will be increased to 16 pounds. If the weight is doubled and the speed remains the same, this force will be doubled. As the radius of the circle described is increased the necessary centripetal force is diminished. Hence, *to produce circular motion, the centripetal force must be increased as the square of the speed increases, and as the mass increases, and diminished as the radius of the circle described increases.*

The farther a point is from the axis of motion,[1] the farther it has to move during a rotation, consequently the greater its speed. Hence, bodies situated at the earth's equator have the greatest speed, due to the earth's rotation, and consequently a greater fraction of the attraction of gravitation is required to keep them moving in the circular path, hence the "apparent weight" is less. Of course the radius of the circle described is greater in this case, but the necessary centripetal force varies directly as the square of the speed, and inversely only as the first power of the radius. It is calculated that a body weighs about $\frac{1}{289}$ less at the equator than at either pole, in consequence of the greater centripetal force at the former place. But 289 is the square of 17; hence, if the earth's rate of rotation were increased seventeen-fold, objects at the equator would weigh nothing.

We have also learned (§ 21) that a body weighs more at the poles in consequence of the oblateness of the earth. This is estimated to make a difference of about $\frac{1}{500}$. Hence, a body will weigh at the equator about $\frac{1}{500} + \frac{1}{289} = \frac{1}{184}$ less than at the poles.

Experiment.—Arrange some kind of rotating apparatus, *e.g.* A, Fig. 31. Suspend a skein of thread *a* by a string, and rotate. Suspend a glass fish aquarium *c*, about one-tenth full of coloured water, and rotate. Pass a string through the

---

[1] *Axis*, an imaginary straight line passing through a body about which it rotates.

longest diameter of an onion *c*, and rotate. State and explain the result in each case.

QUESTIONS

1. Why does not the sphere *d* (Fig. 31) change its position when rotated?
2. State the various facts illustrated in the act of slinging a stone.

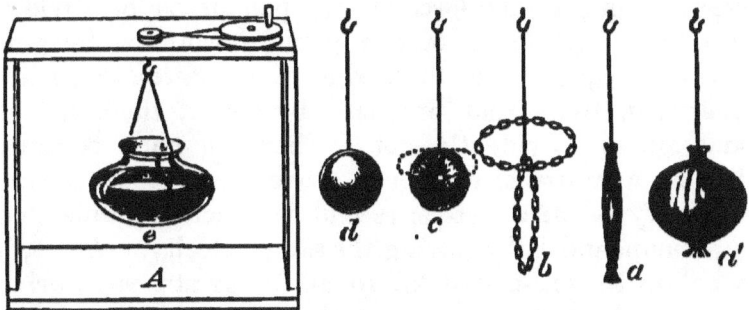

Fig. 31.

3. (*a*) When will water and mud fly off from the surface of a revolving wheel? (*b*) Why do they fly off? (*c*) In what direction do they fly?
4. What is the force that keeps the earth and the other planets in their orbits?
5. How do you account for their curvilinear motion?
6. Understanding that the earth and moon revolve monthly about their common centre of mass, what is the centripetal force acting upon each?
7. How can you account for the high water (tide) on the side of the earth opposite the moon?
8. How can you account for the fact that the equatorial diameter of the earth exceeds the polar diameter?

§ 59. Momentum.—It has already been observed (§ 52) that all bodies, under the action of gravity, fall at the same rate. Thus if a one-pound iron ball and a two-pound iron ball be allowed to fall from rest freely during one second, each at the end of the second has acquired a velocity of 32·2 feet-per-second. Now, on the one-pound ball a force whose intensity is one pound has acted continuously during one second, while on the two-pound ball a force whose intensity is two pounds has acted continuously during one second. Compare the effects. Each force has, during the second, produced the same change of velocity, namely, from rest to 32·2 feet-per-second; but has each produced the same change of motion? In other words, has each produced the same effect? Imagine the force of two pounds to be divided into two equal forces of one pound; can you imagine these two equal forces to produce in one second an effect other than twice that produced by one of them in the same time? Evidently not. Then we must consider that the quantity of motion in a two-pound body having a velocity of 32·2 feet-per-second is exactly double that in a one-pound body having the same velocity. Again, if you allow the one-pound ball to fall from rest freely during two seconds, it, at the end of the two seconds, is found to have a velocity of 64·4 feet-per-second. Now, a one-pound force acting continuously during two seconds must produce just double the effect, in changing motion, of the same force acting continuously during one second. Hence we must conclude that a body of one pound with a velocity of 64·4 feet-per-second has exactly double the quantity of motion possessed by a one-pound body with a velocity of 32·2 feet-per-second. To that which we have called quantity of motion the name *momentum* has been given.

It will be seen from the foregoing that the momentum of a body varies directly as the mass of the body, and also varies directly as the velocity of the body. Hence a proper measure of the momentum of a body is the product of the measure of its mass into the measure of its velocity. The direction of the momentum is, of course, the direction of the velocity. We may also say that *a uniform force acting on a body produces a change of momentum in its own direction proportional to the intensity of the force, and proportional to the time during which the force continuously acts.*

QUESTIONS AND PROBLEMS

1. Compare the momenta of a car weighing 50 tons, moving 10 feet per minute, and a lump of ice weighing 5 cwt., at the end of the third second of its fall.

2. Why are pile-drivers made heavy? Why raised to great heights?

3. A boy weighing 25^k must move with what velocity to have the same momentum that a man has weighing 80^k running at the rate of 10^{km} per hour?

4. A body has a certain momentum after falling through a certain space. How many times this space must it fall to double its momentum?

§ 60. Third Law of Motion.

—It has been shown (§ 50) that motion cannot originate in a single body but arises from mutual action between two bodies. For example, a man can lift himself by pulling on a rope attached to some other object, but not by his boot-straps, or a rope attached to his feet. Whenever one body receives motion, another body always parts with motion, or is set in motion in an opposite direction; that is, *in every change in regard to motion there are always at least two bodies oppositely affected.*

Experiment.—Float two blocks of wood of unequal masses on water, connecting them by a stretched india-rubber band. Let go the blocks. Do both move, or only one? Do they move equally? Float a magnet on a piece of cork and float a piece of iron near it on another piece of cork. What is the result?

A man in a boat weighing one ton pulls at one end of a rope, the other end of which is held by another man, who weighs twice as much as the first man, in a boat weighing two tons: both boats will move towards each other, but in opposite directions; the lighter boat will move twice as fast as the heavier, but with the same momentum.

If the boats are near each other, and the men push each other's boats with oars, the boats will move in opposite directions, though with different velocities, yet with equal momenta.

The opposite impulses received by the bodies concerned are usually distinguished by the terms *action* and *reaction*. We measure these by their momenta. As every force is either a push or a pull (§ 12), and produces equal momenta in two bodies in opposite directions, hence the

THIRD LAW OF MOTION : *To every action there is an equal and opposite reaction.*

The application of this law is not always obvious. Thus, the apple falls to the ground in consequence of the mutual attraction between the apple and the earth. The earth does not appear to fall toward the apple. But, allowing that their momenta are equal, we are not surprised that the motion of the earth is imperceptible, when we reflect that the velocity of the earth must be the same fraction of the velocity of the apple as the mass of the apple is of the mass of the earth (§ 59).

QUESTIONS

1. The velocity of the rebound or "kick" of a gun is slight when compared with the velocity of the ball. Why?
2. In rowing a boat, what are the opposite results of the stress between the oar and the water?
3. Point out the results of the action and reaction that occur when a person leaps from the ground.
4. If there were no ground or other object beneath him, and he were motionless in space, could he put himself in motion? Why?
5. A boy, running, strikes his head against another boy's head. Which is hurt? Why?
6. Suspend two balls of soft putty of equal weight, A and B (Fig. 32). Draw A one side, and let it fall so as to strike B. Both balls will then move on together; with what momentum compared with A's momentum when it strikes B?
7. What will be the momentum of each ball after A strikes B, compared with A's momentum when it strikes B?
8. How will their velocity compare with A's velocity when it strikes B?

9. Raise A and B equal distances in opposite directions, and let fall so as to collide. What is the result? Show that this result is consistent with the third law of motion.

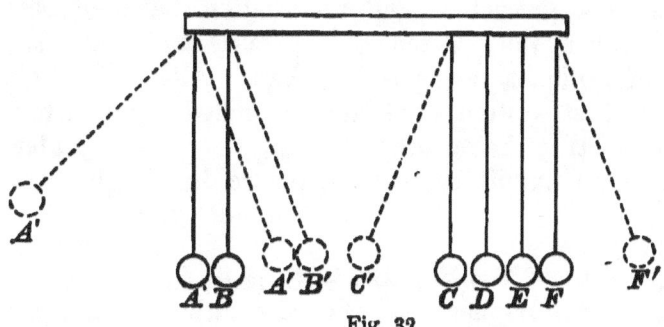

Fig. 32.

10. Substitute for the inelastic putty balls ivory billiard balls, which are highly elastic. Let A strike B. What is the result? Show that this result is consistent with the third law of motion.

11. Suspend four ivory balls, C, D, E, and F. Let C strike D. What happens? Trace the actions and reactions throughout.

12. What would happen if the four balls were inelastic?

Although the laws of motion are clearly indicated by the experiments and observations that have been mentioned, their *proof*, so far as rigorous proof is attainable in physical matters, is furnished in the most conclusive form by observational astronomy. The *Nautical Almanac*, published usually about four years in advance, contains the predicted places of the sun, moon, and principal planets from day to day, in some cases from hour to hour, throughout the year. These predictions are based entirely upon the laws of motion (along with the law of gravitation), and could not be accurate unless these laws are true. The coincidence between prediction and observation has been almost perfect, and variations from absolute coincidence have been found to be due not to errors in the statement of the laws, but to errors in our estimates of the masses of the planets, or of their orbits. Indeed, in the brilliant investigations of Adams and Leverrier these slight variations from absolute coincidence between prediction and observation have enabled us to discover the existence, and even to assign the position of a planet never before seen. Although these laws were first systematised and extended by Newton, after whom they have been named, they were understood to some extent some time before by Galileo and others.

§ 61. **Law of Reflection.**—Experiment 1.—Hold D (Fig. 32) firmly in its place, and allow C to strike it. D being immovable, C's entire momentum is spent in compressing the balls, and, on recovering their shape, C is thrown back to its starting-point at C'. But in this case the hand exerts as much force to prevent the motion of D as would be necessary to project C to C'. *When an elastic body strikes another fixed perfectly elastic body, it rebounds with its original momentum.*

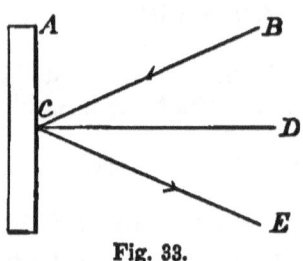

Fig. 33.

Experiment 2.—Lay a marble slab A (Fig. 33) upon a table, and roll an ivory ball in the line DC, perpendicular to the surface of the slab; the ball rebounds in the same line to D. Roll the ball in the line BC; it rebounds in the line CE. The angle BCD, which its forward path makes with DC, a perpendicular to the surface struck, is called the *angle of incidence*. The angle ECD, which its retreating path makes with the same perpendicular, is called the *angle of reflection*.

It is found by measurement that these angles are nearly equal when the two bodies are very elastic. This equality is expressed by the LAW OF REFLECTION : *When the striking body and the body struck are perfectly elastic, the angle of reflection is equal to the angle of incidence.*

By resolving the velocity of the ball into two components, one at right angles to the fixed plane and the other parallel to it, show that this law is a necessary consequence of the laws of motion.

§ 62. **Work.**—A force may produce either motion or pressure (or tension), or it may produce both effects at the same time and in the same body. But *a force does work,* in the sense in which this term is used in science, *only when it produces motion in a body.* The body that is moved is said to have work done upon it ; and the body that moves another body is said to do work upon the latter. When the heavy weight of a pile-driver is raised, work is done upon it ; when it descends and drives the pile into the earth, work is done upon the pile, and the pile in turn does work upon the matter in its path.

Whenever a force causes motion, it does work. A force may act for an indefinite time without doing any work ; but *whenever a force acts on a body through space, work is done.* Force and space (or distance) are essential elements of work, and are naturally the quantities employed in estimating work. A given force producing motion through a space of one metre will do a certain amount of work ; it is evident that the same force producing motion through a space of two metres will do twice as much work. Hence the general formula,

$$W = FS \qquad (1),$$

in which W represents the work done, F the force employed, and S the space through which the force produces motion.

In case a force encounters resistance, the magnitude of the force necessary to produce motion depends upon the amount of resistance. Indeed, in cases in which the body having been moved through a given space comes to rest in consequence of resistance, the entire work done upon the body is often more conveniently determined by *multiplying the measure of the resistance by the measure of the space through which it is overcome,* and our formula becomes by substitution of resistance, R, for the force which overcomes it,

$$W = RS \qquad (2).$$

For example, a ball is shot vertically upward from a rifle in a vacuum ; the work done upon the ball may be estimated by multiplying the measure of the average force (difficult to ascertain) exerted upon it by the measure of the space through which this force acts (a little greater than the length of the barrel), or by multiplying the measure of the resistance offered by gravity, *i.e.* its weight (easily ascertained) by the measure of the distance the ball ascends. Also, in case the motion produced is uniform, the resistance and the force are equal, and it is immaterial which formula is used. When there is no resistance and the only effect is acceleration, as when a body falls freely in a vacuum, we must estimate the work done (in this case by gravity) by the first formula. When it is required to estimate only that part of the work done in producing acceleration, the formulas given in § 68 will be found convenient, work being substituted for energy, inasmuch as both are measured by the same units.

§ 63. Unit of Work.—We shall first consider the unit employed when resistance is taken as one of the elements of work. The unit of work adopted by the French is the work done in raising 1^k through a vertical height of 1^m against the action of gravity. It is called a *kilogrammeter* (abbreviated ^{kgm}). The English unit of work is that done in raising one pound one foot, and is called a *foot-pound*. The kilogrammeter is about $7\frac{1}{4}$ (more accurately, 7·233) times the foot-pound. Now, since the work done in raising 1^k 1^m high is 1^{kgm}, the work of raising it 10^m high is 10^{kgm}, which is the same as the work done in raising 10^k 1^m high; and the same, again, as raising 2^k 5^m high.

There are many other kinds of work besides that of raising weights. But since, with the same resistance, the work of producing motion in any other direction is just the same as in a vertical direction, it is easy, in all cases in which the two elements of work (viz. resistance and space) are known, to find the equivalent in work done in raising a weight vertically. By thus securing a common standard for measurement of work, we are able to compare any species of work with any other. For instance, let us compare the work done by a man in sawing through a stick of wood, whose saw must move 10^m against an average resistance of 12^k, with that done by a bullet in penetrating a plank to a depth of 2^{cm} against an average resistance of 200^k. Moving a saw 10^m against 12^k resistance is equivalent to raising 12^k 10^m high, or doing 120^{kgm} of work; a bullet moving 2^{cm} against 200^k resistance does as much work as is required to raise 200^k 2^{cm} high, or $200 \times ·02 = 4^{kgm}$ of work. $120 \div 4 = 30$ times as much work done by the sawyer as by the bullet.

§ 64. **Rate of doing Work.**—In estimating the total amount of work done, the time consumed is not taken into consideration. The work done by a hod-carrier, in carrying 1000 bricks to the top of a building, is the same whether he does it in a day or a week. But in estimating the *power* of any agent to do work, as of a man, a horse, or a steam-engine, in other words, *the rate* at which it is capable of doing work, it is evident that time is an important element. The work done by a horse, in raising a barrel of flour 20 feet high, is about 4000 ft.-lbs.; but even a mouse could do the same amount of work in time. The unit in which rate of doing work is usually expressed is a *horse-power*. Early tests showed that a very strong horse may perform 33,000 ft.-lbs. of work in one minute. So 1 horse-power = 33,000 ft.-lbs. per minute = 550 ft.-lbs. per second = about 4570^{kgm} per minute = about 76^{kgm} per second.

§ 65. **Energy.**—The *energy* of a body is its capacity of doing work, and is measured by the work it can do. *Doing work consists in a transfer of energy from the body doing work to the body on which work is done.* Wherever we find matter in motion, whether in the solid, liquid, or gaseous state, we have a certain amount of energy which may often be made to do useful work.

§ 66. **Potential and Kinetic Energy.**—Place a stone, weighing (say) 10^k, on the floor before you; it is devoid of energy, powerless to do work. Now raise it, and place it on a shelf (say) 2^m high; in so doing you perform 20^{kgm} of work on it. As you look at it, lying motionless on the shelf, it appears as devoid of energy as when lying on the floor. Attach one end of a cord 3^m long to it, and, passing it over a pulley, wind 2^m of the string around the shaft connected with a sewing-machine, coffee-mill, lathe, or other convenient machine. Suddenly withdraw the shelf from beneath the stone. It moves, it sets in motion the machine, and you may sew, grind coffee, turn wood, etc., with the energy given to the machine by the stone.

Surely the work done on the stone in raising it was not lost; the stone pays it back while descending. There is a very important difference between the stone lying on the floor, and the stone lying on the shelf: the former is powerless to do work; the latter can do work. Both are alike motionless, and you can see no difference, except an *advantage* that the latter has over the former *in position.* What gave it this advantage? Work. *A body then, may possess energy due merely to* ADVANTAGE OF POSITION, *derived always from work done upon it.* So a body at rest is not necessarily devoid of energy. In the stone lying passively on the shelf there exists a power to do work as real as that possessed by the stone which, falling freely, has acquired great speed.

We see, then, that energy may exist in either of two widely different states, and yet be as real in one case as in the other. It may exist as *actual motion*, either visible, as in mechanical motion, or invisible, as in the molecular motions called heat; or it may exist in a *stored-up condition*, as in the stone lying on the shelf. In the former case it is called *kinetic* (moving) or *actual* energy; in the latter, it is called *potential* energy, or *energy of position.*

We are as much accustomed to store up energy for future use as to store up provisions for the winter's consumption. We store it when we wind up the spring or weight of a clock, to be doled out gradually in driving the machinery. We store it when we bend the bow, raise the hammer, condense air, or raise any body above the earth's surface.

How, then, is energy stored in a body? *Only at the expense of work done upon it.* The force of gravitation is employed to do useful work, as when mills are driven by the energy of falling water; but the water is first deposited on the hillside by the energy of the sun's heat. Elasticity of springs is employed as a motive power; but elasticity is due to an advantage of position which the molecules of springs have acquired in consequence of force applied to them, so as to move them away from the positions which their mutual attractions have a tendency to make them occupy.

We conclude, then, that *a body possesses potential energy when, in virtue of work done upon it, it occupies a position of advantage, or its molecules occupy positions of advantage, so that the energy expended can be at any time recovered by the return of the body to its original position, or by the return of its molecules to their original positions.*

§ 67. **Energy contrasted with Momentum.**—Problem.—A bullet weighing 30^g is shot with a speed of 98^m per second from a gun weighing 4^k; required the momentum and the energy of both the bullet and the gun, and the speed of the gun. *Solution:* Using the kilogram, the metre, and the second as units, the momentum of the ball is $\cdot 03 \times 98 = 2\cdot 94$ units. If the ball were shot vertically upward, its speed would diminish $9\cdot 8^m$ per second; so it would rise $\dfrac{98}{9\cdot 8} = 10$ seconds, and, therefore, before its energy is expended, to a height of (§ 54) $4\cdot 9^m \times 10^2 = 490^m$. Hence, its energy at the outset is $\cdot 03 \times 490 = 14\cdot 7^{kgm}$. Similarly for the gun, by the third law of motion, its momentum must be just the same as that of the ball, $2\cdot 94$ units; its speed is therefore $2\cdot 94 \div 4 = \cdot 735^m$ per second. Then $T = \dfrac{\cdot 735}{9\cdot 8} = \cdot 075$ second; the height (supposing the gun to be raised vertically by the impulse received) $= 4\cdot 9 \times \cdot 075^2 = \cdot 02766^m$; and its energy $= 4 \times \cdot 02766 = \cdot 1102^{kgm}$.

While, therefore, the momenta generated in the two bodies by the burning of the powder are equal, the energy of the bullet is $\dfrac{14\cdot 7}{\cdot 1102} = 133\frac{1}{3}$ times that of the gun. (Why are the effects produced by the bullet more disastrous than those produced by the recoil of the gun?)

§ 68. **Formula for Energy.**—We can find, as in the above example, to what vertical height a body having a given speed would rise, and thus in all cases determine its energy; but a formula may be obtained which will give the same result with less trouble; thus:—

Let E = measure of energy, in foot-pounds.
„ V = „ speed, in feet-per-second.
„ g = „ acceleration due to gravity = 32·2.
„ S = „ height to which body would rise in feet.
„ W = „ weight of body, in pounds.
„ T = „ time of rising, in seconds.
„ V = gT

$$\therefore T = \frac{V}{g}$$

or $T^2 = \dfrac{V^2}{g^2}$

$S = \tfrac{1}{2} g T^2$

$= \tfrac{1}{2} g \times \dfrac{V^2}{g^2}$

But $S = \dfrac{V^2}{2g}$

$E = WS$

$= \dfrac{WV^2}{2g}$

It is evident that, *when the weight* (W) *of a body remains the same, its energy is proportional to the square of its speed, while its momentum,* as we have learned, *is proportional to its speed.* In other words, the effect of increasing the speed of a moving body would seem to be to increase its working power much more rapidly than its momentum. Is this *practically* true?

Experiment.—Fill an ordinary water-pail with moist clay. Let a leaden bullet drop upon the clay from a height of ·5^m. Then drop the same bullet from a height of 2^m, or four times the former height, in order that it may acquire *twice* the speed. Carefully measure the depth to which it penetrates in each case. How does the work done by the bullet in the first case compare with the work done by it in the second case?

So it appears that *the energy of a moving body varies,* not as its speed, but *as the square of its speed.* Doubling the speed multiplies the energy fourfold; trebling the speed multiplies it ninefold, and so on; but the corresponding momentum is multiplied only twofold, threefold, etc. A bullet moving with a speed of 400 feet-per-second, will penetrate, not twice, but four times, as far into a plank as one having a speed of 200 feet-per-second. A railway train, having a speed of 20 miles an hour, will, if the steam is shut off, continue to run four times as far as it would if its speed were 10 miles an hour. The reason is now apparent why light substances, even so light as air, exhibit great energy when their speed is great.

§ 69. **Measure of a Force.**—Commonly we measure forces by a spring balance, and say that the force, for instance, with which a horse draws a wagon is 50^k; that is, a spring interposed between the horse and the wagon is stretched just as much as it would be by the force of gravity acting on a mass of 50^k hung from the spring. But often it is impossible to measure the force except by the motion it produces. Experience has shown that *a useful and accurate measure of a force is the momentum it produces or destroys in a second;* if the body is already in motion, we must say the *change of momentum produced in a second.*

For example, gravity we know will (at the surface of the earth in the latitude of London) impart in three seconds, to a body having a mass of (say) 5^g, and free to fall, a velocity of $3 \times 981\cdot2^{cm}$ per second; that is, the momentum generated is measured by $5 \times 3 \times 981\cdot2$. Then, by definition above, the measure of the force of gravity on the body is $\frac{5 \times 3 \times 980}{3} = 5 \times 981\cdot2$. When the centimeter, gramme, and second are taken as the units of length, mass, and time respectively, the system of units of measurement based on them is called the C.G.S. system, and in it the unit of force is called a *dyne.*

A dyne is that force which, acting for a second, will give to a gramme of matter a velocity of one centimeter-per-second. In the example above we have a force of $(981\cdot2 \times 5 = 4906)$ dynes.

The *gravity unit of force* is the weight of any unit of mass, *e.g.* a gramme, kilogramme, pound, or ton. In distinction from gravity units, the C.G.S. units are called *absolute units*. Gravity units are easily changed to absolute units; thus in Great Britain the force of gravity acting upon 1^g of matter free to fall will give it an acceleration of velocity of $981·2^{cm}$ per-second per second; hence, in these latitudes, the gravity unit is equal[a] to 981·2 absolute units.

Let M = measure of mass of body, in grammes.
,, W =   ,,   weight of body, in dynes.
,, F =   ,,   attraction between earth and body, in dynes.
,, g =   ,,   acceleration due to gravity, in absolute units = 981·2.

$$W = F = Mg.$$

The equation is a general one; that is, whenever any two of the three quantities specified are known, the third may be computed.

If the force acts, not against gravity, but against resistances considered as constant, such as the forces shown in cohesion, elasticity, etc., the equation will still be true, only g should be replaced by some other letter, as a.

It will be observed that we have different ways of measuring a force.

1. We may observe the mass which the force will support against the action of gravity.

2. We may observe the change of momentum produced by the force in a known time. Considered in this light a force may be said to be the *time rate of change of momentum*.

3. We may observe the work done by the force in producing motion through a known space. Considered in this light a force may be said to be the *space rate of transference of energy*.

The first of these methods is subject to the objection

that the action of gravity is not the same at all points on the earth's surface.

§ 70. Summary of mechanical Units, and Formulas for their Determination.[1]—The following tables show the quantities measured, the unit of each in the C.G.S. system, and the formulas for the determination of the derived quantities :—

FUNDAMENTAL QUANTITIES AND UNITS

Length (L or S) . . 1^{cm}
Mass (M) . . . 1^g
Time (T) . . . 1 sec.

DERIVED QUANTITIES, UNITS, AND FORMULAS

Velocity (V) = rate of motion ; unit, 1^{cm} per sec. ; in uniform motion, $V = \dfrac{S}{T}$. (1)

Acceleration (A) = rate of change in velocity ; unit, an increase of velocity in 1 sec. of 1^{cm} per sec. ; body starting from rest under constant acceleration, $A = \dfrac{V}{T}$ (2)

Force (F) ; unit, 1 dyne = a force that in 1 sec. imparts to 1^g a velocity of 1^{cm} per sec. ; ∴ $F = MA$. (3)

Work or Energy (E) ; unit, 1 erg = the work done by 1 dyne producing motion through 1^{cm} ; ∴ $E = MAS = FS$. (4)

Rate of doing work, or Work Power (P) ; unit, 1 erg per sec. ; $P = \dfrac{MAS}{T}$. (5)

Momentum ; unit, 1^g moving with a velocity of 1^{cm} per sec. or that produced by 1 dyne in 1 sec. ; Momentum = MV.

From (2) and (3) we have the very useful equations, $F = \dfrac{MV}{T}$ $V = \dfrac{FT}{M}$. (6) and (7)

---

[1] It is not expected that pupils of the ordinary schools will master this section; yet they may frequently find it convenient for reference, while the more advanced student cannot fail to be greatly profited by its careful study.

A body, mass M, acted upon by the force F, starting from rest will acquire in time T a velocity $V = \dfrac{FT}{M}$. The acceleration, which from (3) is $= \dfrac{F}{M}$, is a constant quantity, and the whole space passed over will be equal to the time T multiplied by the *mean* velocity. The latter is one-half the final velocity; hence, mean $V = \dfrac{FT}{2M}$, and $S = \dfrac{FT^2}{2M}$ (an equation of great importance). (8)

To find an expression for the energy of a moving body combine (4) and (8): $W = \dfrac{F^2 T^2}{2M}$; but $FT = MV$, $\therefore E = \dfrac{MV^2}{2}$. (9)

Anywhere in Great Britain, the weight of $1^g = 981\cdot 2$ dynes approximately.

$1^{kgm} = 98,120,000$ ergs; 1 foot-pound $= 13,567,000$ ergs.
1 horse-power $= 447,711,000,000$ ergs per min.

§ 71. Transformation of Energy.

—In the operation of raising the stone (§ 66), kinetic energy is transformed into potential energy. During its descent it is re-transformed into kinetic energy. If, instead of being attached to machinery, and thereby made to do work, the stone is allowed to fall freely, it acquires great velocity. On striking the ground, its motion as a body suddenly ceases, but its molecules have their quivering motions accelerated. Mechanical motion is, thereby, transformed into heat. We shall often have occasion to examine the transformations of energy, as into electric energy, heat, etc., but never of momentum. We shall study Joule's equivalent (§ 135), expressing the relation between the unit of energy, or work, and the unit of heat.

§ 72. Physics defined.

—All physical phenomena consist either alone in transferences of energy from one portion of matter to another, or in both transferences and transformations of energy. Transformations may be from one condition of energy to another, as from kinetic to potential; or from one phase of kinetic energy to another, as from mechanical motion to heat; or both may occur, as when the falling stone does work, a part of its energy being expended in producing mechanical motion, and a part being transformed into heat, occasioned by friction of the moving parts.

Physics is that branch of natural science which treats of transferences and transformations of energy. It does not, however, in its usual limitation, include a group of phenomena which occur outside the earth, and also a group whose essential characteristic is an alteration in the nature of the material considered. The study of the former group is the object of *Astronomy*; of the latter, that of *Chemistry*.

QUESTIONS AND PROBLEMS

1. Does the energy expended in raising the stones to their places in the Egyptian pyramids still survive?

2. What kind of energy is that contained in gunpowder?

3. What transformation of energy takes place in burning coal?

4. When steam works by expansion, its temperature is reduced. Why?

5. How much work is done per hour if 80^k are raised 4^m per minute?

6. (*a*) What energy must be imparted to a body weighing 50^g that it may rise 4 seconds? (*b*) How many times as much energy must be imparted to the same body that it may ascend 5 seconds? (*c*) Why?

7. Compare the momenta, in the two cases given in the last question, at the instants the body is thrown.

8. How much energy is stored in a body which weighs 50^k, and is resting at a height of 80^m above the earth's surface?

9. How much energy would the same body have if it had a velocity of 100^m per second?

10. Suppose it to fall from rest in a vacuum, how much kinetic energy would it have at the end of the fourth second?

11. If it should fall through the air, what would become of a part of the potential energy lost by it in falling?

12. A projectile weighing 25^k is thrown vertically upward with an initial speed of $29 \cdot 4^m$ per second. How much energy has it?

13. What becomes of its kinetic energy during its ascent?

14. (*a*) Compare the momentum of a body weighing 50^k, and having a velocity of 2^m per second, with the momentum of a body weighing 50^g, having a velocity of 100^m per second. (*b*) Compare their energies.

15. Which, momentum or energy, will enable one to determine the amount of resistance that a moving body may overcome?

16. Explain how a child who cannot lift 30^k can draw a carriage weighing 150^k.

17. A carriage weighing 6000^k is drawn by a horse with a speed of 100^m per minute. The index of the dynamometer to which the horse is attached stands at 40^k. (*a*) At what rate is the horse working? (*b*) Express the rate in horse-powers. (See § 64.)

18. A dynamometer shows that a span of horses pull a plough with a constant force of 70^k. What power is required to work the plough if they travel at the rate of 3^{km} per hour?

19. What horse-power in an engine will raise $1,350,000^k$ 5^m in an hour?

20. How long will it take a 3 horse-power engine to raise 10 tons 50 feet?

21. How far will a 2 horse-power engine raise 1000^k in 10 seconds?

22. How much work can a 5 horse-power engine do in an hour?

23. How long would it take a man to do the same work, the amount of work a man can do in a day being about $90,000^{kgm}$?

24. If you would increase the energy of a moving body fourfold, how much must you increase its speed?

§ 73. The Pendulum.—Experiment 1.—From a bracket suspend by very fine strings leaden balls, as in Fig. 34. Draw B and C one side, and to different heights, so that B may swing through a short arc, and let both drop at the same instant. Count the number of vibrations of B in five

minutes, while another counts the number of vibrations of C. Repeat the experiment several times, varying the lengths of the arcs through which the pendulums swing. What general conclusion is indicated by these experiments?

Experiment 2.—Make B 1^m long, measured from the centre of the ball, and make F $\frac{1}{4}^m$ long. Compare the number of vibrations made by F in five minutes with the number made by B in the same time. Make G $\frac{1}{9}$ the length of B, and count the number of vibrations made by G in five minutes. From a study of these observations what conclusion do you reach regarding the relation existing between the length of a pendulum and its period of vibration? From the number of vibrations made by B in five minutes find the period of one vibration, and from this and the law just discovered calculate the length of a pendulum vibrating once in a second.

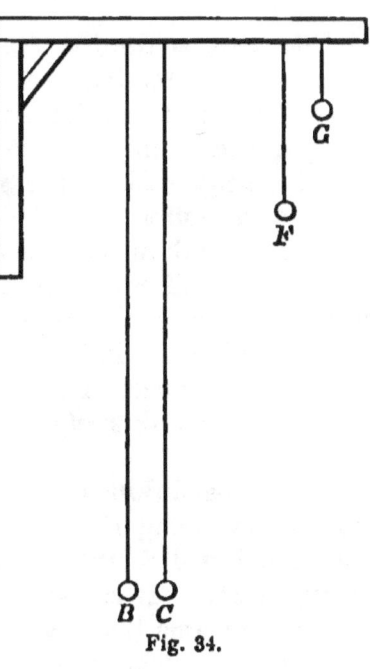

Fig. 34.

QUESTIONS AND PROBLEMS

1. What would be the effect if B were made twice as heavy as C? Why?

2. What is the length of a pendulum that beats half-seconds? Quarter-seconds? That makes one vibration in two seconds? That makes two vibrations per minute?

3. State the proportion that will give the number of vibrations per minute made by a pendulum 40^{cm} long.

4. What *transformations* of energy are going on in the pendulum bob as it swings? Is there any *transference* of energy from the pendulum bob to anything else? To what?

5. If the pendulum were swinging in a vacuum would there be any transference of energy? In this case what must be the

speed of the bob at the lowest point of its swing, supposing this point to be one foot nearer the earth than the highest point of its swing ?

§ 74. Some useful Applications of the Pendulum.—The force that keeps a pendulum vibrating is gravity. Were it not for friction and the resistance of the air, a pendulum, once set in motion, would never cease vibrating. Since the force of gravity keeps the pendulum in motion, it follows that the rate of vibration of a given pendulum must be determined by the intensity of this force. Hence it is obvious that, if the rate of vibration is known, the intensity of the force of gravity may be calculated. It is found by experiment, and is a necessary consequence of the laws of motion, that *the time of vibration varies inversely as the square root of the force of gravity.*

Thus the pendulum becomes a most serviceable instrument for measuring the intensity of gravity at various altitudes and at different latitudes on the earth's surface. (Compare § 21.) It is also the most accurate instrument for measuring time that has been invented. Its value, as a time-measurer, depends upon the absolute uniformity of the rate of vibration so long as its length is constant, and the length of its arc very small. But as heat is ever modifying the dimensions of bodies, various devices have been called into existence by which heat may be made to correct automatically its own mischief. Clocks that do not have self-regulating pendulums are fast in winter and slow in summer. (How would you regulate them ?)

STATICS

We shall next proceed to investigate the relations that must exist among a number of forces acting upon a body in order that the motion of this body may remain unchanged ; that is, in order that these forces may counterbalance one another. If a system of forces counterbalance one another, the system is said to be in *equilibrium*.

§ 75. Composition of Forces.—It is often desirable to know what single force will produce the same effect as two or more given forces produce when acting upon a body simultaneously. The force to be found in this case is called the *resultant* of the given forces, and the problem of finding it is called *composition of forces*.

By Experiment 3 (§ 52) it is demonstrated that—*If two forces acting at a point are represented in magnitude and direction (on the same scale) by two straight lines drawn from that point, and the parallelogram be constructed having these lines as adjacent sides, then that diagonal of this parallelogram which passes through to this point represents in magnitude and direction (still on the same scale) the resultant of the two given forces.*

The foregoing proposition is known as *the parallelogram of forces*, and is very important.

Let two forces of 8 lbs. and 4 lbs. acting at A (Fig. 35) be represented in magnitude and direction by AB and by AD, then, by the parallelogram of forces, AC represents the resultant of these forces in magnitude and direction.

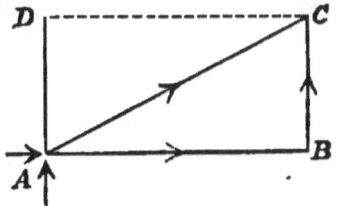

Fig. 35.

The numerical value of the resultant may be found by comparing the length of the line AC with the length of either AB or AD, whose numerical values are known. Thus, AC is 2·23 times AD ; hence the numerical value of the resultant AC is $4 \times 2·23 = 8·92$.

When the components act at right angles to each other, as in Fig. 35, the line representing the resultant divides the parallelogram into two equal right-angled triangles ; and the intensity of the resultant may be found by calculating the hypothenuse, having two sides of either triangle given.

Thus, $\sqrt{4^2+8^2}=8\cdot 9+$ is the numerical value of the resultant AC. When the two components of a force F act at right angles to each other, each component is called the *resolved part* of the force F in the direction of that component. Since no force has any effect in a direction at right angles to its line of action, it is evident that the *resolved part* of a given force in any direction is the *effective component* of the given force in that direction.

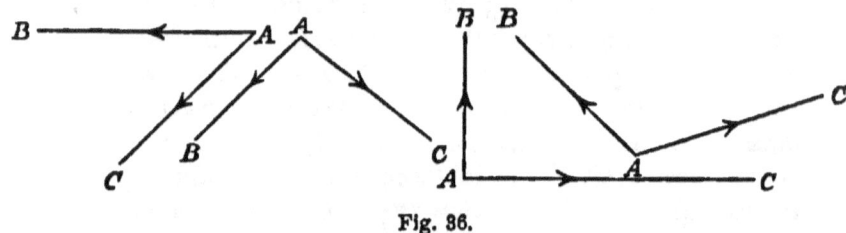

Fig. 36.

Copy upon paper and find the resultant of the components AB and AC, in each of the four diagrams in Fig. 36. Also assign appropriate numerical values to each component, and find the corresponding numerical value of each resultant.

§ 76. Finding the Resolved Parts of Forces.— For the sake of brevity we will call a force represented

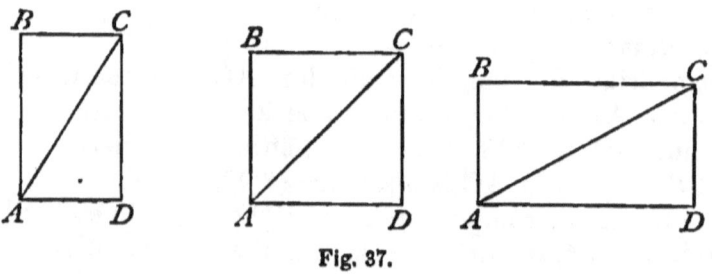

Fig. 37.

in intensity by the length of AC and in direction by the direction of AC, the force AC. Let DB be a rectangle

(Fig. 37). From what has been said it will appear that the forces AB and AD are the resolved parts of the force AC in the directions of AB and of AD respectively. Now, if AC is given, and the angle CAB is given, AB and AD may be found. However, as we wish to deal only with very elementary mathematics, we shall consider only those cases in which the angle CAB is 30°, 45°, or 60°.

If the angle CAB is 30°, the triangle ACB is evidently half an equilateral triangle (Euclid I. 5 and 32), and $BC = \frac{1}{2}$ AC, and therefore $AB = \frac{\sqrt{3}}{2}$ AC (Euclid I. 47). If the angle CAB is 45°, $AB = BC$ (Euclid I. 32 and 6), and therefore $AB = \frac{\sqrt{2}}{2}$ AC (Euclid I. 47). If the angle CAB is 60°, the angle ACB is 30°, and therefore AB is $\frac{1}{2}$ AC.

From the above, of the truth of which the student should thoroughly satisfy himself, it will be seen *that the resolved part of a force F in a direction making an angle of 30° with its line of action is* $F \times \frac{\sqrt{3}}{2}$; *if the angle is 45° the resolved part is* $F \times \frac{\sqrt{2}}{2}$; *if the angle is 60°, the resolved part is* $F \times \frac{1}{2}$; *and if the angle is 90°, the resolved part is zero.* These four facts should be carefully remembered, as they will be found exceedingly useful.

QUESTIONS

1. Find the vertical and horizontal resolved parts of a force of 100 lbs., whose line of action makes an angle of 60° with the vertical.

2. A picture weighs 20 lbs., and the two parts of the suspension-cord make with each other at the nail an angle of 60°; what is the tension of the cord?

3. If a body is in contact with a *smooth* surface, why is the

pressure between the body and surface at *right angles* to the surface ?

4. Why is the pressure between a fluid and a surface in contact with it at right angles to the surface ?

5. Why can a sailing yacht make progress in a direction at right angles to the direction in which the wind is blowing ? In what position must its sail be set ?

6. What is the force which raises a kite ?

§ 77. Composition of Forces by first finding Resolved Parts.

Let two forces of 14 lbs. and 10 lbs.

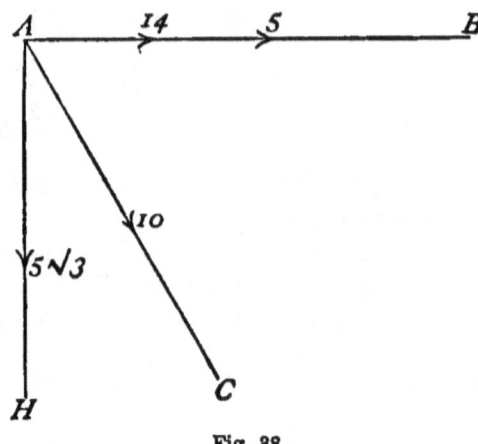

Fig. 38.

act along the lines AB and AC (Fig. 38), and let the angle BAC = 60°. Find the resultant of these two forces.

Draw AH at right angles to AB.

For the force of 10 lbs., substitute its resolved parts along AB and AH.

Since the angle BAC = 60°, and angle CAH = 30°, these resolved parts are 5 lbs. along AB, and $5\sqrt{3}$ lbs. along AH.

Hence the original forces have the same effect as a force of 19 lbs. along AB, and a force of $5\sqrt{3}$ lbs. along AH.

Now, since AB and AH are at right angles, the resultant of these forces $= \sqrt{(19)^2 + (5 \times \sqrt{3})^2}$ lbs.

Let two forces of 6 lbs. and 4 lbs. act along the lines AB and AC (Fig. 39), and let the angle BAC = 75°. Find the resultant of these two forces.

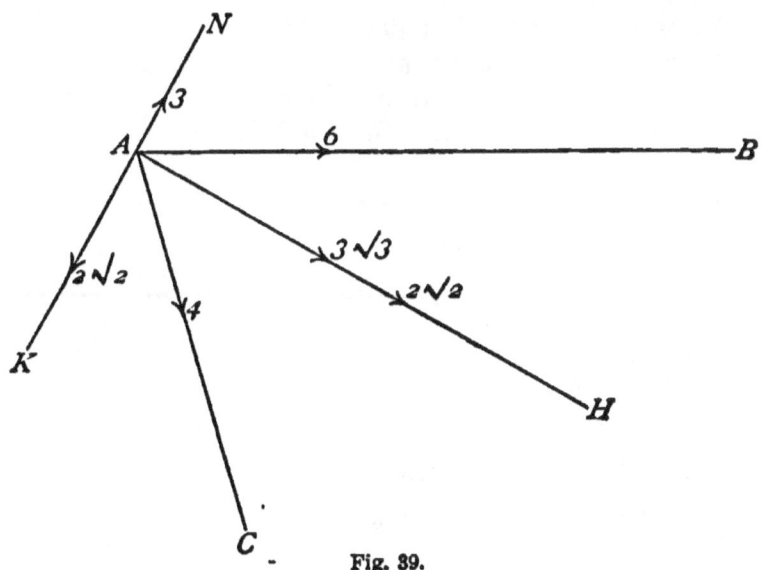

Fig. 39.

Draw AH, making the angle BAH = 30°, and draw NAK at right angles to AH.

Now it is easily seen that—

The angle BAN = 60°;
,, ,, CAH = 45°;
,, ,, CAK = 45°.

For the force of 6 lbs., acting along AB, substitute its resolved parts, 3 lbs., along AN and $3\sqrt{3}$ lbs. along AH.

For the force of 4 lbs. along AC, substitute its resolved parts, $2\sqrt{2}$ lbs. along AH, and $2\sqrt{2}$ lbs. along AK.

Hence the original forces have the same effect as $(3\sqrt{3} + 2\sqrt{2})$ lbs. along AH, and $(3 - 2\sqrt{2})$ lbs. along AN.

Now, since AH and AN are at right angles, the

resultant of these forces $= \sqrt{(3\sqrt{3}+2\sqrt{2})^2 + (3-2\sqrt{2})^2}$ lbs.

Let three forces of 8 lbs., 5 lbs., and 6 lbs., act along AB, AC, and AD (Fig. 40). Let the angle BAC = 60°, and let the angle CAD = 60°. Find the resultant of these three forces.

Draw HAK at right angles to AC.

Now it is easily seen that—

The angle BAH = 30°;
,, ,, DAK = 30°.

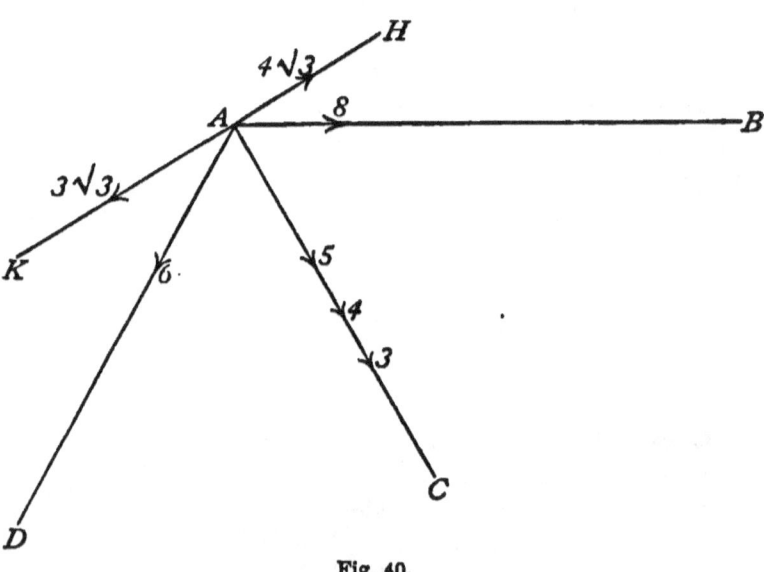

Fig. 40.

For the force of 8 lbs. acting along AB, substitute its resolved parts, 4 lbs., along AC, and $4\sqrt{3}$ lbs. along AH.

For the force of 6 lbs. acting along AD, substitute its resolved parts, 3 lbs., along AC, and $3\sqrt{3}$ lbs. along AK. Hence the original three forces have the same effect as a force of $(5+4+3)$ lbs. = 12 lbs. along AC, and $(4\sqrt{3} - 3\sqrt{3})$ lbs. = $\sqrt{3}$ lbs. along AH.

Now, since AC and AH are at right angles, the re-

sultant of these forces $= \sqrt{(12)^2 + (\sqrt{3})^2}$ lbs.
$= \sqrt{147}$ lbs.

In a similar manner may be found the resultants of other systems of forces.

QUESTIONS

1. Two forces each of 50 lbs. act at a point at an angle of 60°; find the resultant.
2. Forces of 12 and 20 lbs. respectively act at a point at an angle of 60°; find the resultant.
3. Forces of 8 and 10 lbs. respectively act upon a particle at an angle of 30°; find the resultant.
4. Two forces each of 6 lbs. act at a point at an angle of 45°; find the resultant.
5. Forces of 10 and $12\sqrt{3}$ lbs. respectively act at a point at an angle of 150°; find the resultant.
6. Find the resultant of two forces of 8 and 10 lbs. acting at a point at an angle of 15°.
7. Two forces of 15 and 40 lbs. respectively act at a point at an angle of 120°; find their resultant.
8. Two forces acting at an angle of 135° have a resultant of $10\sqrt{5}$ lbs.; one of the forces is 30 lbs.; find the other.
9. Two forces of 10 and 12 lbs. act at a point at an angle of 75°; find their resultant.
10. Two equal forces act at an angle of 120°; their resultant is 10 lbs.; find the forces.
11. If a force of 60 lbs. be resolved into two equal forces acting at an angle of 30°, what is the magnitude of either component?
12. The resultant of two equal forces acting at an angle of 135° is 20 lbs.; find the components.
13. The resultant of two forces in the ratio of 2 : 3, acting at an angle of 150°, is 24 lbs.; find the forces.
14. The resultant of two forces acting at an angle of 60° is 21 lbs.; one of the components is 9 lbs.; find the other.

§ 78. Composition of Parallel Forces.—If two parallel forces act at the same point and in the same direction, it is obvious that this resultant has that direction, and is equal to the sum of the given forces. If two parallel forces act at the same point and in opposite directions, it is equally obvious that this resultant has the direction of the greater of the given forces, and that it is equal to the difference of the given forces. When parallel forces are not applied at the same point, the question arises, What is the point of application of their resultant? To the opposite extremities (Fig. 41) of a bar AB, having very little weight (Why?), apply two sets of weights, which shall be to each other as 3 : 1. The resultant is a single force, applied at some point between A and B. To find this point it is only necessary to find a point where a single force must be applied to prevent motion resulting from the parallel forces; in other words, to find a point where a support may be applied so that the whole will be balanced. Use a spring-balance to support the bar, and carefully ascertain the point C about which the bar balances. Observe the force indicated by the spring-balance, and measure AC and CB. What is the force indicated by the spring-balance? What relation does it bear to the weights suspended from A and B? What is the ratio of AC to CB? What connection do you observe between this ratio and the ratio of the weight suspended from A to the weight suspended from B? What is the direction of the force acting upon the bar at C?

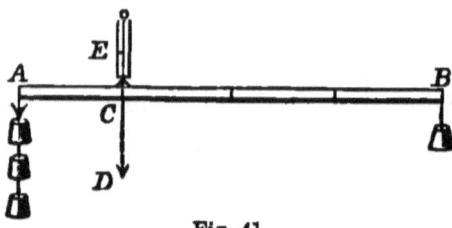

Fig. 41.

Repeat the experiment several times, using different weights, being careful always to use weights great when compared with the weight of the bar (§ 1). What general conclusion do you reach (1) regarding the *magnitude* of the resultant of two parallel forces acting at different points and in the same direction; (2) regarding the *direction* of this resultant; (3) regarding the *point of application* of this resultant?

Regarding the force acting at B as equal and opposite to the resultant of the forces acting at A and at C, what do you conclude regarding the magnitude, direction, and point of application of the resultant of two parallel forces acting at different points and in contrary directions?

§ 79. Couple.—If two *equal, parallel,* and *opposite* forces are applied to opposite extremities of a stick AB (Fig. 42), no single force can be applied so as to keep the stick from moving; there will be no motion of *translation,* but simply a *rotation* around its middle point C. Such a pair of forces, *equal, parallel,* and *opposite,* but *not in the same line,* is called a *couple.*

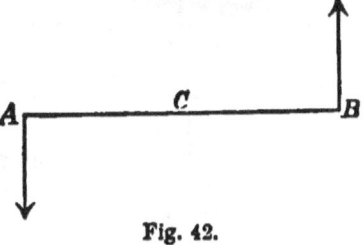

Fig. 42.

PROBLEMS, ETC.

1. A man and a boy, grasping opposite ends of a pole 3^m long, support thereon a weight of 50^k between them. Where should the weight be placed that the boy may support 20^k?

2. If the weight were placed 40^{cm} from the man, how much would each support?

3. Suppose that a boat is headed directly across a river half a mile wide, and is rowed with a velocity that would land it upon the opposite shore in half an hour, if there were no current; but the current carries the boat down the stream at the rate of one mile an hour. Where will the boat land?

4. How far will it travel?

5. How long will it be in crossing the river?

6. A ship is sailing due south-east at the rate of ten miles per hour; what is its southerly velocity?

7. Find, both by construction and by calculation, the intensity of two equal forces, acting at right angles to each other, that will support a weight of 15 lbs.

8. Verify the results with dynamometers.

§ **80. Centre of Gravity.**—Let Fig. 43 represent any body of matter; for instance, a stone. Every molecule of the body is acted upon by the force of gravity; the intensity of this force is measured by the *weight* of the molecule. The forces of gravity of all the molecules form a set of parallel forces acting vertically downward, the resultant of which equals their sum, and has the same direction as its components. The resultant can be proved to have a definite point of application in whatever position the body may be, and this point is called its *centre of gravity*. *The centre of gravity (c.g.) of a body is, therefore, the point of application of the resultant of all these forces;* and for many purposes *the whole weight of the body may be supposed to be concentrated at its centre of gravity.* Hence mathematicians, by the *place* of a body, usually mean that point where the c.g. is situated.

Fig. 43.

Let G in the figure represent this point. For many practical purposes, then, we may consider that gravity acts only upon this point, and in the direction GF. If the stone falls freely, this point cannot, in obedience to the first law of motion, deviate from a vertical path, however much the body may rotate during its fall. Inasmuch, then, as the c.g. of a falling body always describes a definite path, a line GF that represents this path, or the path in which a body supported tends to move under the action of gravity, is called the *line of direction.* The centre of gravity coincides, in some bodies absolutely, and in all bodies approximately, with the *centre of inertia,* the point at which the whole *mass* of a body may for certain purposes be supposed to be concentrated.

It is evident that if a force equal to its own weight and opposite in direction is applied to a body anywhere in the line of direction (or its continuation), this force will be the equilibrant of the forces of gravity; in other words, the body subjected to such a force is in equilibrium, and is said to be *supported*, and *the equilibrant* is called its *supporting force*. *To support any body*, then, *it is only necessary to provide a support for its centre of gravity.* The supporting force must be applied somewhere in the line of direction, otherwise the body will fall.

Experiment.—Place a stick of wood, two metres long, horizontally across the tip end of a finger. When you succeed in getting the finger directly under its c.g., it will rest, but not till then. The difficulty of poising a book, or any other object, on the end of a finger, consists wholly in keeping the support under the centre of gravity.

Fig. 44 represents a toy called a "witch," consisting of a cylinder of pith terminating in a hemisphere of lead. The toy will not lie in the position shown in the figure on a horizontal surface *ab*, because the support is not applied immediately under its c.g. at G; but, when placed

Fig. 44.

horizontally, it immediately assumes a vertical position. It appears to the observer to rise; but, regarded in a mechanical sense, it really falls, because its c.g., where all the weight may be supposed to be concentrated, takes a lower position.

Whether a body will stand or fall depends upon whether or not its line of direction falls within its base. The base of a body is not necessarily limited to that part of the under surface of a body that touches its support. For example, place a string around the four legs of a table close to the floor: the rectangular figure bounded by the string is the base of the table. (What is the base of a man when standing on one foot? on two feet?)

§ 81. **How to find the Centre of Gravity of a Body.**—**Experiment.**—Attach a string to a potato by means of a tack, as in Fig. 45, and suspend from the hand. When the potato comes to rest there will be an equilibrium of forces, and the c.g. must be in the same line with the equilibrant of gravity; hence, if a knitting-needle is thrust vertically through the potato from a, so as to represent a continuation of the vertical line oa, the c.g. must lie somewhere in the path an made by the needle. Suspend the potato from some other point, as b, and a needle thrust vertically through the potato from b will also pass through the c.g. Since the c.g. lies in both the lines an and bs, it must be at c, their point of intersection. Find by trial the centre of gravity of a triangle. Cut out several triangles from cardboard, and from your experiments deduce a general conclusion.

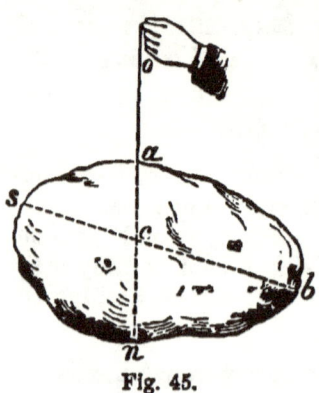

Fig. 45.

§ 82. Three States of Equilibrium.—The weight of a body is a force tending downward; hence, *a body tends to assume a position such that its c.g. will be as low as possible.*

Experiment 1.—Try to support a ring on the end of a stick, as at b (Fig. 46). If you can keep the support exactly under the c.g. of the ring, there will be an equilibrium of forces, and the ring will remain at rest. But if it is slightly disturbed, the equilibrium will be destroyed, and the ring will fall. Support it at a. Have you any difficulty in supporting it in this position? Disturb the ring and remove the disturbing force. What happens? Why?

A body is said to be in *stable* equilibrium if its position is such that a very slight disturbance would raise its c.g., since in that event it would tend to return to its original position. On the other hand, a body is said to be in *unstable* equilibrium when a very slight disturbance would lower its c.g., since it would not, on the disturbing cause being removed, return to its original position.

A body is said to be in *neutral* or *indifferent* equilibrium when it rests equally well in any position in which it may be placed. A sphere of uniform density, resting on a horizontal plane, is in neutral equilibrium, because its c.g. is neither raised nor lowered by a change of base. Likewise, when the support is applied at the c.g., as when a wheel is supported by an axle, the body is in neutral equilibrium.

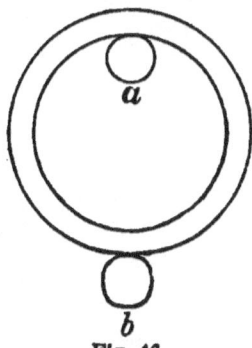

Fig. 46.

It is evident that, *if the c.g. is vertically below the support*, as in the last experiment with the ring, *the equilibrium must be stable;* but, as in Fig. 44, a body may be in stable equilibrium though its c.g. is above the point of support. (When is this possible ?)

It is difficult to balance a lead-pencil on the end of a finger; but by attaching two knives to it, as in Fig. 47, it may be rocked to and fro without falling. Explain.

Fig. 47.

§ 83. **Stability of Bodies.**—The ease or difficulty with which a body supported at its base is overturned depends upon the height to which its c.g. must be raised in overturning it, and upon the weight of the body. What is a measure of the *work* that is done in overturning a body? The letter c (Fig. 48) marks the position of the c.g. of each of the four bodies A, B, C, and D. To turn any one of these bodies over, its c.g. must pass through the arc ci, and be raised through the height ai. By comparing A with B, and supposing them to be of equal weight, we learn that *of two bodies of equal height and weight, the c.g. of that body which has the larger base must be raised higher, and this body is, therefore, overturned with greater difficulty.* A comparison of A and

C, supposing them to be of equal weight, shows that *when two bodies have equal bases and weights, the higher body is more easily overturned.* D and C have equal bases and heights, but D is made heavy at the bottom, and this *lowers its c.g. and gives it greater stability.*

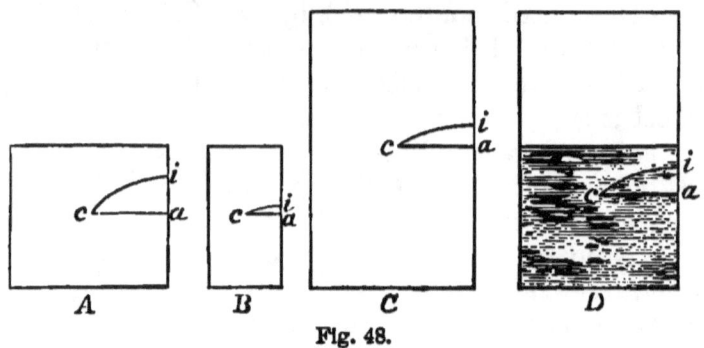

Fig. 48.

QUESTIONS

1. Where is the c.g. of a box ?
2. Why is a pyramid a very stable structure ?
3. What is the object of ballast in a vessel ?
4. State several ways of giving stability to an inkstand.
5. (a) In what position would you place a cone on a horizontal plane, that it may be in stable equilibrium ? (b) That it may be in neutral equilibrium ? (c) That it may be in unstable equilibrium ?
6. In loading a waggon, where should the heavy luggage be placed ? Why ?
7. Why are bipeds slower in learning to walk than quadrupeds ?
8. How will a man rising in a boat affect its stability ?
9. Which is more liable to be overturned, a load of hay or a load of stone of equal weight ? Why ?
10. Place a book upon the table in different positions, so that it may have three different degrees of stability, and account for these differences by considering the energy that must be expended in each case to overturn the book.
11. Doors and gates are often made self-closing by using hinges of peculiar pattern, and without the use of any spring. How is this done ? What physical law is taken advantage of ?

§ 84. **Moments.**—It is often convenient to consider the tendency of a force to produce rotation about a given point. This tendency is called the *moment* of the force with respect to that point. For the investigation of moments make use of the apparatus described in § 78, Fig. 41. When the bar AB is balanced about the point C it is obvious that, neglecting the weight of the bar, which we may do provided the weight of the bar is *very small when compared with the weights suspended at A and B*, the moment about the point C of the weight suspended at A is equal and opposite to the moment about C of the weight suspended at B. Make several experiments, and determine the relation between the moment of a given force about a given point and the magnitude of the given force; also determine the relation between the moment and the perpendicular from the given point on the line of action of the given force. These relations are easily inferred. Test your inference by many experiments. To use arithmetic or algebra in making calculations regarding moments we must use the *measures* of the moments. To obtain the measure of a moment we must adopt a unit of moment. The most convenient unit of moment is found to be the moment of the unit of force about a point the unit distance from its line of action. For example, if one pound is the unit force and one foot the unit distance, the unit moment used is the moment of a force of one pound about a point one foot from its line of action. In terms of this unit what is the measure of the moment of a force of 10 lbs. about a point 5 feet from its line of action?

When two moments are to be combined it must be observed whether they have the same or opposite signs; that is, whether the tendency to rotation is in the same direction in both cases or not. For convenience a tendency to rotation in the direction opposite that in which the hands of a watch move when looking at its face is considered positive.

The moment of a couple (§ 79) about any point is, of course, the resultant moment about that point of the two forces constituting the couple. Determine the moment of a particular couple about different points. What do you find ?

§ 85. Equilibrium of Forces acting in the same Plane.

—Let us investigate the conditions necessary in order that a system of forces acting in the same plane may be in equilibrium ; that is to say, in order that this system of forces acting together on a body may have no tendency to give it motion either of *translation* or of *rotation*. We may reach general conclusions by carefully considering a few questions. What is the moment of a force about a point in its line of action ? Is there any point outside its line of action about which its moment is the same ? If the moment of a force P about O is equal and opposite to the moment of a force Q about O, what conclusion may you draw concerning the line of action of the resultant of P and Q ?

When, in a system of forces, the moments about a point O of those forces of the system which tend to produce rotation about that point in one direction, are together equal to the moments about O, of those forces of the system which tend to produce rotation about that point in the opposite direction, the moments of the whole system are said to *vanish about the point* O.

If the moments of a system vanish about a point O, what do you know about the line of action of the resultant of that system ?

If the moments of the system vanish about O, and also vanish about P, what do you know about the resultant of that system ? If the moments vanish about each of three points O, P, and Q, what do you know about the resultant ? If O, P, and Q are not in the same straight line, what do you know about the resultant ? Can the line of action of the resultant of a system of forces be other than a straight line? If a system of forces is in equilibrium, that is, if the forces are such as to counteract one another, and, taken together, produce no effect, either of translation or of rotation, is there any point about which their moments do not vanish?

If the resolved parts (§ 76) of a system of forces along a line in one direction are together equal to the resolved parts of the same system along the same line in the opposite direction, the resolved parts of that system are said to *vanish along that line*.

If the resolved parts of a system of forces vanish along the line AB, can that system, as a whole, produce any motion along AB in either direction? In the above case, what do you know about the resultant of the system?

If the resolved parts of the system vanish along AB, and also along CD, what do you know about the resultant? If AB and CD are not parallel, what do you know about the resultant? Can the line of action of the resultant be at right angles to each of two lines which are not parallel to each other? If a system of forces is in equilibrium, is there any line along which their resolved parts do not vanish? If the resolved parts of a system vanish along each of two lines not parallel to each other, is the system necessarily in equilibrium? Is a couple (§ 79) in equilibrium? Do the resolved parts of a couple vanish along each of two lines not parallel?

From a careful consideration of the foregoing questions the pupil will see the truth of the following propositions, which are very important:—

EQUILIBRIUM OF FORCES ACTING IN THE SAME PLANE

1. *If a system of forces is in equilibrium the resolved parts of the system vanish along any line whatever, and the moments of the system vanish about any point whatever.*

2. *If the moments of a system of forces, all in the same plane, vanish about each of three points not in the same straight line, the system must be in equilibrium.*

3. *If the moments of a system of forces, all in the same plane, vanish about one point, and their resolved parts vanish along each of two straight lines not parallel to each other, the system must be in equilibrium.*

If it is known that a system of forces is in equilibrium, the first proposition may be made use of to form equations involving the intensities and lines of action of the various forces, from which equations such intensities and lines of action as are not known may be determined.

If the forces of the system all act in the same plane not more than three *independent* equations can be based upon the fact that the system is in equilibrium. Why ?

Example 1.—A uniform beam, AB, 20 feet long, weighing

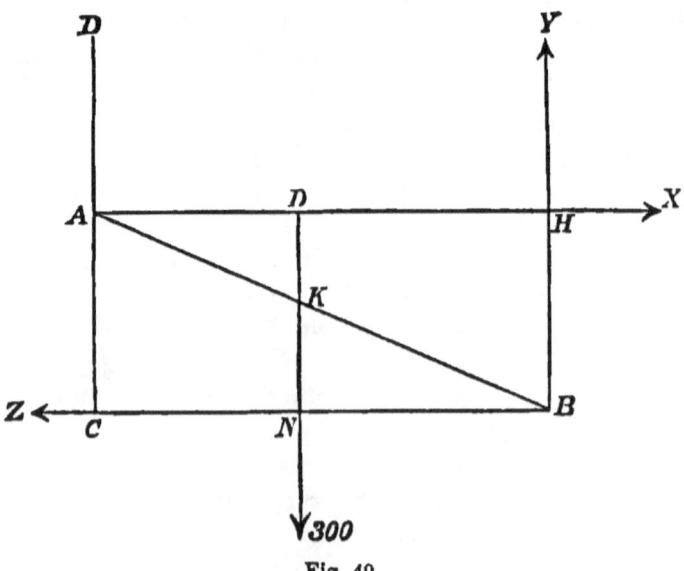

Fig. 49.

300 lbs., rests with one end against a smooth, vertical wall, CD, and the other end on a smooth, horizontal plane, CB,

this end being tied by a cord, CB, 16 feet long. Investigate the forces acting upon the beam.

The beam is evidently acted upon by four forces, namely: 1st, its own weight, 300 lbs., which, since the beam is uniform, may be supposed to act at K, the middle point of AB; 2d, the tension of the string CB, z lbs., acting at B in the direction BC; 3d, the pressure of the wall, x lbs., acting at A at right angles to the wall, since the wall is smooth; 4th, the pressure of the floor, y lbs., acting at B at right angles to the floor since the floor is smooth.

The beam and the forces acting upon it are represented in Fig. 49.

Since the beam is at rest, the forces acting upon it are in equilibrium; therefore, applying the first proposition above, we have the following equations:—

Because the resolved parts of the system vanish along a horizontal line,
$$x = z \qquad . \qquad . \qquad . \qquad . \qquad (1).$$
Because the resolved parts vanish along a vertical line,
$$y = 300 \qquad . \qquad . \qquad . \qquad . \qquad (2).$$
Because the moments vanish about any point (say) H,
$$z\,BH = 300\,OH \qquad . \qquad . \qquad . \qquad (3).$$

Now, since ABH is a right-angled triangle, and AB is twenty feet, and $AH = CB = 16$ feet; therefore, $BH = 12$ feet (Euclid I. 47).

Therefore, substituting in equation (3) we have
$$12\,z = 300 \times 8$$
$$z = \frac{300 \times 8}{12} = 200.$$
But $x = z$,
$$\therefore x = 200.$$

Example 2.—Let AB (Fig. 50) be a smooth inclined plane, the angle A being 30°. Let a heavy particle placed at D be kept at rest by a string, DB. Investigate the forces acting on this particle. The particle is evidently acted on by three forces, namely: 1st, its own weight, 100 lbs., acting like all weights vertically downward; 2d, the tension of the string, x lbs., acting in line DB; 3d, the pressure of the plane, y lbs., acting at D, and, since the plane is smooth, acting at right angles to the plane.

Since the particle D is at rest, the forces acting on it are in equilibrium; therefore, applying the same proposition as before, we have the following equations:—

Because the resolved parts vanish along the line AB and the angle ADE = 60°,

$$x = 100 \times \tfrac{1}{2} = 50.$$

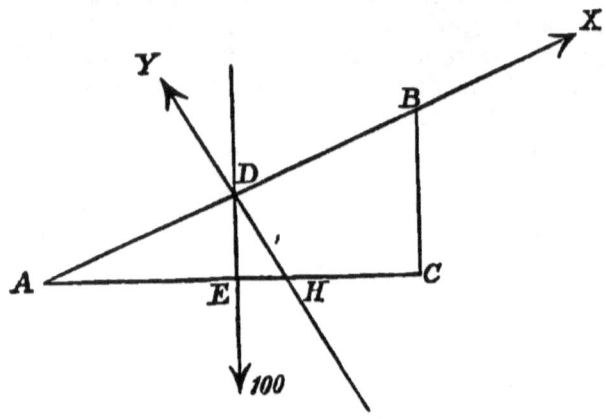

Fig. 50.

Because the resolved parts vanish along DH, and the angle EDH = 30°,

$$y = 100 \times \frac{\sqrt{3}}{2} = 50\sqrt{3}.$$

Example 3.—Four weights of 5, 7, 9, and 11 lbs. are placed at the corners of a uniform square plate, weighing 12 lbs., whose side is 10 inches; find the distance of the c.g. from the centre of the plate.

Imagine the plate to be placed in a horizontal plane, as in Fig. 51. Now all the weights are forces acting at right angles to the plate. Let O be the centre of the plate. Let G be the c.g. of the plate and weights taken together. Therefore a force of (12 + 5 + 7 + 9 + 11) lbs., or 44 lbs., acting through G is the resultant of the five weights acting through A, B, C, D, and O. Hence the moment about any line of 44 lbs. acting at G is equal to the sum of the moments about the same line of the five weights.

From O and from G let fall perpendiculars on AD and DC. Let GF = x inches, and let GK = y inches.

Equating moments about AD, we obtain the equation
$$44\,x = (12 \times 5) + (7 \times 10) + (9 \times 10),$$
$$\therefore x = 5.$$
Equating moments about DC, we obtain the equation
$$44\,y = (12 \times 5) + (5 \times 10) + (7 \times 10),$$
$$\therefore y = 4\tfrac{1}{11}.$$
Since G is 5 inches from AD and $4\tfrac{1}{11}$ inches from DC, it is evidently $\tfrac{10}{11}$ inch from the centre of the plate.

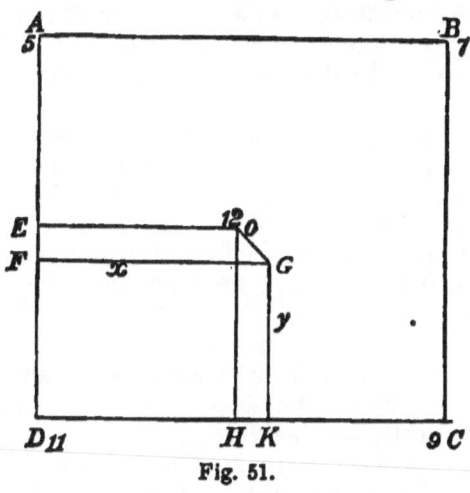

Fig. 51.

To apply the foregoing propositions correctly, the pupil must be careful to take note of *all* the forces of the system in equilibrium, and to make no mistake in expressing the resolved parts and moments. Resolved parts are discussed in §§ 75 and 76.

QUESTIONS AND PROBLEMS

1. If you wish to base an equation on the fact that the moments of a system of forces in equilibrium vanish about any point whatever, and wish the equation not to involve a particular force of the system, where should you choose the point? If you wish the equation not to involve either of two forces of the system, what point must you choose? If you wish an equation, based on the fact that the resolved parts of the system vanish along any line whatever, not to involve a particular force of the system, what line should you choose?

2. A uniform beam 37 feet long, weighing 400 lbs., rests with one end against a smooth wall, and the other end on a smooth floor, this end being fastened by a cord 12 feet long to a peg at the bottom of the wall; find the tension of the cord.

3. In Example 2, if the weight of the beam acted at a point $\frac{1}{3}$ of its length from the lower end; find the reactions of the wall and floor.

4. A beam AB, weighing 120 lbs., acting at its middle point, is made to rest against a smooth vertical wall and on a smooth floor, by a force applied horizontally to the foot; find the force if the inclination of the beam is (a) 30°, (b) 45°, (c) 60°.

5. In Example 4, if a weight of 90 lbs. was suspended on the beam at a point (a) $\frac{1}{3}$ of its length from the foot, (b) $\frac{3}{4}$ of its length from the foot; find the horizontal force, the inclination of the beam being (a) 30°, (b) 45°, (c) 60°.

6. In Example 4, if the weight of the beam acted at a point $\frac{2}{3}$ of its length from the foot; find the force, the inclination of the beam being (a) 30°, (b) 45°, (c) 60°.

7. Taking the inclination of the beam as in Example 4, find where a weight of 120 lbs. must be suspended so that the horizontal force may be the same as that in Example 5, the beam being uniform, and weighing 120 lbs.

8. A uniform beam, weighing 60 lbs., rests with one end against a peg in a smooth horizontal plane, and the other end on a wall. The point of contact with the wall divides the beam into parts, as 3 : 8; find the pressure on the peg, and the reaction of the wall, the inclination of the beam being (a) 30°, (b) 45°, (c) 60°.

9. What relation must the moment of the whole weight of a body about any line bear to the algebraic sum of the moments of the weights of its several parts about the same line?

N.B.—By algebraic sum of the moments of several forces about a point or line is understood, the excess of the sum of the positive moments over the sum of the negative moments. It is customary to consider a moment positive when the tendency to rotation is in the direction opposite that in which the hands of a watch move when you are looking at its face, and negative, of course, when the tendency to rotation is in the same direction as the hands of the watch.

10. How may you apply the answer to Question 9 to

find the distance of the centre of gravity of a body from a given line, when the weights of the several parts of the body are given, and the distance of the centre of gravity of each part from that line is also given?

N.B.—The weight of a body may be supposed to act at its centre of gravity (§ 80).

11. How may you apply the answer to Question 10 to find the position of the centre of gravity of the body in Question 10?

12. Three uniform rods are placed so as to form a right-angled isosceles triangle, the longest being $8\sqrt{2}$ feet; find their c.g.

13. A heavy, tapering rod, 10 feet long, weighing 160 lbs., balances about a point 4 feet from the heavy end, when a weight of 30 lbs. is attached to the other end; find the point about which it will balance if the 30 lbs. is removed.

14. Four weights of 5, 7, 9, and 11 lbs. are placed at the corners of a uniform square plate weighing 12 lbs., whose side is 10 inches; find the distance of the c.g. from the centre of the plate.

15. Weights of 4, 6, 8, 7, 3, 2, 13, and 1 lbs. are placed at the corners and middle points of the sides of a square taken in order; find their c.g., the side of the square being 16 inches.

16. A uniform square plate whose side is 10 inches, and weight 12 lbs., has a weight of 18 lbs. attached to one corner; where must it be suspended by a cord so as to rest horizontally?

17. A square table, weighing 40 lbs., rests on four legs, one at each corner. Can you determine the pressure on each leg? If so, how?

18. From a circular plate whose radius is 8 inches, a circular plate whose radius is 4 inches is cut away, the distance between the two centres is 2 inches; find the centre of gravity of the remainder.

19. From a uniform circular disc, whose diameter is 10 inches, another disc, having for its diameter the radius of the first circle, is cut away; find centre of gravity of the remainder.

20. From a circular disc whose diameter is D, a circular disc whose diameter is d is cut away; find the centre of gravity of the remainder, the distance between their centres being a.

§ 86. Uses of Machines.—Experiment 1.—Obtain from an ironmonger two or three pulleys, and arrange ap-

paratus as in Fig. 52. If the power applied to maintain rest in each instance is slightly increased, the weights will rise. Raise each of the weights and measure the distances traversed respectively by W and P in each. Through what distance must P move that W may be raised one foot? What amount of work is done in raising 8 lbs. one foot? How does this work compare with the work done at P? Is any work saved

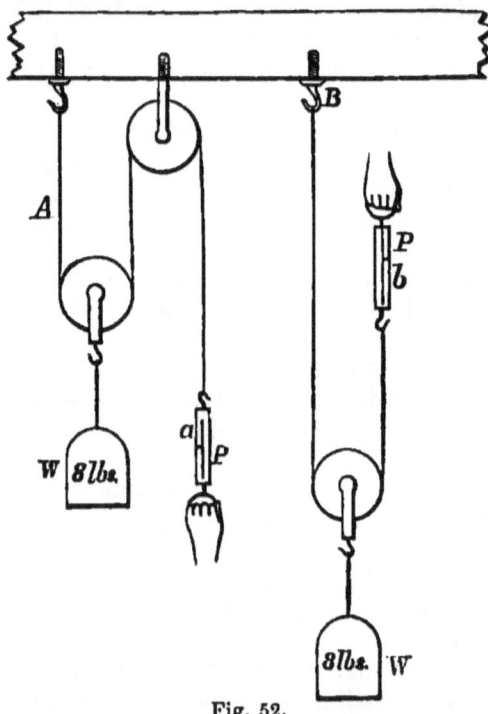

Fig. 52.

by raising the weight by means of the pulley instead of lifting it directly? Is a force of less intensity required when the pulley is used? Since the string is light, and passes round freely revolving pulleys, its tension may be supposed to be the same throughout its length. What is its tension? Since the 8 lbs. is supported by two parallel portions of the string, what must the tension be? What does the dynamometer P show the tension to be?[1]

[1] A small allowance must be made for the weight of the movable pulleys.

Experiment 2.—Let P and W of A exchange places. What does the index of the dynamometer now read? Allow motion to take place, and note carefully the height W is raised while P descends one foot. In this case how does the work done at W compare with that done at P? What is the tension of the string? Is there any advantage gained in this case by the use of apparatus?

This apparatus is one of many *contrivances called machines, through the mediation of which power may be applied to resistance more advantageously than when it is applied directly to the resistance.* Some of the many advantages derived from the use of machines are:—

(1) *They may enable us to overcome a resistance of great intensity with a power of little intensity by causing the power to move through a proportionately greater distance than that through which the resistance is moved; or, conversely, they may enable us to secure great velocity by employing a power of proportionately greater intensity than the resistance.*

(2) *They may enable us to employ a force in a direction that is more convenient than the direction in which the resistance is to be moved.*

(3) *They may enable us to employ other forces than our own in doing work;* e.g. the strength of animals, the forces of wind, water, steam, etc. (How are the last two uses illustrated in Fig. 53?)

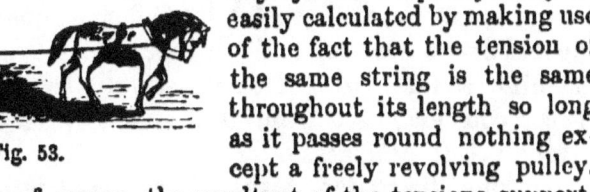
Fig. 53.

The ratio of the weight to the power in any system of pulleys may be easily calculated by making use of the fact that the tension of the same string is the same throughout its length so long as it passes round nothing except a freely revolving pulley. The weight is, of course, the resultant of the tensions supporting it, and the power is the tension or the resultant of the tensions by which it is supported. What is the ratio of the weight to the power in Fig. 53?

§ 87. Law of Machines.—Let P be the power

applied to a machine, p the distance through which it moves in its own direction in a given time, W the weight moved or external resistance overcome, and w the distance through which it is moved in its own direction in the same time; then the mechanical work applied to the machine is Pp (*e.g.* in kilogrammeters or foot-pounds), and the mechanical work done by the machine is Ww. Now we have learned from the foregoing experiments that

(1) $Pp = Ww$.

It may be shown that for all machines, without exception, the following general law holds true: *The work applied to a machine is equal to the work done by the machine.*

No machine, therefore, creates or increases energy. No machine gives back more energy than is spent upon it. P can be made as small as we please by taking p great enough: in this case we see that *in proportion as intensity of force is gained, time, distance, or velocity is lost.* On the other hand, W remaining the same, w (the distance traversed by W in a given time, *i.e.* its velocity) may be increased indefinitely by taking P large enough: in this case, *as velocity, time, or space is gained, intensity of force is lost.* A machine, then, is much like a bank: it pays out no more than it receives. A bank will give you in exchange for a fifty pound note fifty sovereigns; or, for fifty sovereigns, deposited successively, it will return to you a fifty-pound note. In a similar manner, if you apply to a machine energy sufficient to move 50 lbs. 1 ft., you may get from it the ability to move 1 lb. 50 ft.; or, if you apply to a machine a force of 1 lb. successively through 50 ft. of space, you may get from it the ability to move 50 lbs through 1 ft. of space.

In our discussion hitherto we have ignored the internal resistances, chiefly due to friction, which exist in every machine. The whole work done by a machine is practically divided into two parts,—the *useful* part and the *wasted* part; the former, expressed as a fraction of the whole, is usually called the *efficiency* or *modulus* of the machine. But energy is indestructible. That portion of the visible energy that is apparently destroyed by friction is transformed into heat, which is wasted, so far as the work to be done by the machine is concerned. Let I represent *internal work* performed *in* the machine, *i.e.* the *wasted work*, and Ww the external work; then our general formula for machines, as modified in its practical applications, becomes

$$(2) \quad Pp = Ww + I \,;$$

that is, *the work applied to a machine is equal to the effective work, plus the internal work done by the machine.* So, that so far from any machine being a *source of energy*, as is sometimes erroneously supposed, no machine practically returns as much useful energy as is applied to it.

By division, Formula (1) $Pp = Ww$ becomes

$$(3) \quad \frac{W}{P} = \frac{p}{w},$$

i.e. *weight : power : : the distance through which the power moves : the distance through which the weight is moved in the same time.* Problems pertaining to machines may generally be solved by Formula (3), and afterwards suitable allowances may be made for the internal work done. Thus, suppose that P (Fig. 54) is 10 lbs., and it is required to find what weight (W) it will raise. Find by experiment or by geometry how far P moves (say x feet) in its direction, while W moves 1 foot in its direction.

Thus $\quad Px = W,$

$$P = \frac{w}{x}.$$

If the lever is at rest under the influence of the forces acting upon it, we may apply the propositions regarding forces in equilibrium (§ 85). Thus the moments of these forces about any point (say b) must vanish.

$$\therefore Pab = Wcb,$$

$$\therefore P = W\frac{cb}{ab}.$$

K

The many attempts that have been made to obtain "perpetual motion" have assisted greatly in the discovery of that most important physical law that, *by no contrivance whatever is it possible to create or to annihilate energy.*

It is to be observed that, as we saw (§ 63), work is not

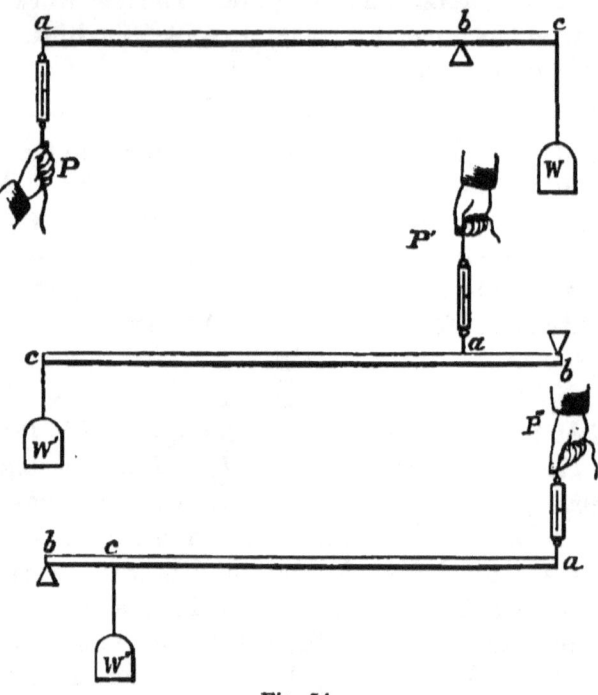

Fig. 54.

always, or even usually, expended in raising a weight, but in overcoming resistance of any kind; so we may interpret Formula (3) thus : *resistance : power : : the distance through which the power moves : the distance through which the resistance is overcome.*

QUESTIONS AND PROBLEMS

1. If the power applied to any machine is 2^k, and it moves with a velocity of 10^m per second, with what velocity can it

move a resistance of 10^k? To how great a resistance could it give a velocity of 50^m per second?

2. A power of 50^k, moving through a space of 100^m, is capable of raising how many kilogrammes through a space of 2^m? What advantage would be gained by the use of the machine?

3. Watch the movements of the foot in working the treadle of a sewing-machine, also the movements of the needle in sewing, and determine what mechanical advantage is gained by the machine.

4. Arrange three levers, as in Fig. 54; and, calling the distance (ab) of the power from the prop the *power-arm* of the lever, and the distance (bc) of the weight from the prop the *weight-arm*, verify by experiment the following special formula for levers:—

$$\frac{W}{P} = \frac{p}{w} = \frac{\text{power-arm}}{\text{weight-arm}}.$$

N.B.—The weight of the lever itself should be small compared with P and W. Why? See (§ 1) advice regarding making experiments.

5. Ascertain the advantage that may be gained by each lever.

6. A lever is 75^{cm} long; where must the prop be placed in

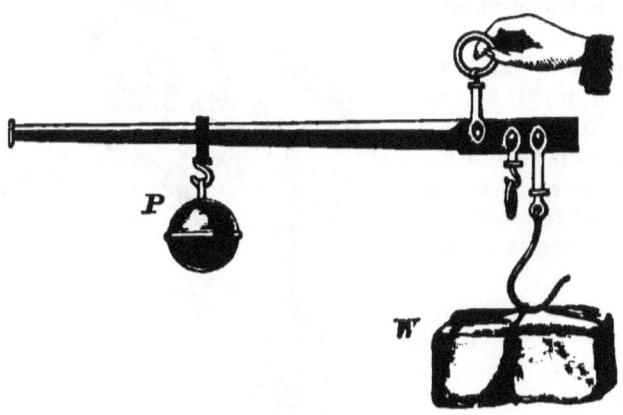

Fig. 55.

order that a power of 2^k at one end may move 4^k at the other end? What will be the pressure on the prop?

7. Show that the results obtained in the last problem are consistent with what you have learned about parallel forces (§ 78).

8. What advantage is gained by a lever, when its power-arm is longer than its weight-arm? What, when its weight-arm is longer?

Fig. 56.

9. Two weights, of 5^k and 20^k, are suspended from the ends of a lever 70^{cm} long. Where must the prop be placed that they may balance?

10. What mechanical advantage is gained by a lemon-squeezer?

11. If P (Fig. 55), weighing 1 lb., is suspended 15 spaces from the fulcrum of the steelyard, what weight (W) suspended three similar spaces the other side of the fulcrum will balance it, provided that the empty beam balances about the fulcrum?

12. How would you weigh out 6 lbs. of tea with the same steelyard?

13. If the circumference of the axle (Fig. 56) is 60^{cm}, and the power applied to the crank travels 240^{cm} during each revolution, what power will be necessary to raise the bucket of coal weighing (say) 40^k?

14. How many metres must the power travel (Fig. 56) to raise the bucket from a cavity 10^m deep?

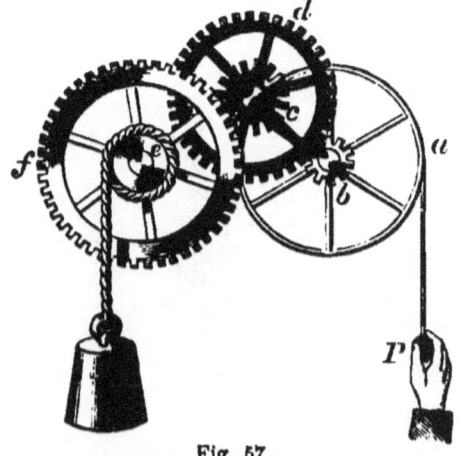

Fig. 57.

15. (a) In the train of wheels (Fig. 57), if the circumference of the wheel a is 36 inches, and that of the pinion b is 4 inches, a power of 1 lb. at P

will exert what force on the circumference of the wheel d ? (*b*) If the circumference of the wheel d be 30 inches, and that of the pinion c 6 inches, the power of 1 lb. at P will exert what force on the circumference of the wheel f ? (*c*) If the circumference of the wheel f be 40 inches, and that of the axle e 8 inches, how many pounds in W will be necessary to prevent motion of the train of wheels, when P weighs 1 lb. ? (*d*) If W has a velocity of 5 feet per second, what will be P's velocity ?

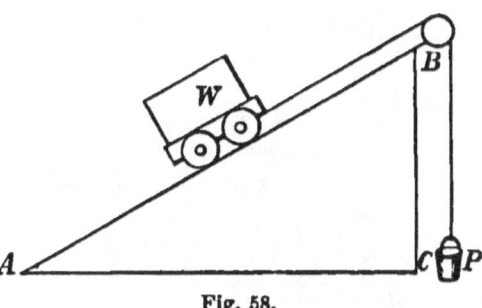

Fig. 58.

16. Prepare a special formula for the solution of problems pertaining to the wheel and axle.

17. The weight W (Fig. 58), in traversing the inclined plane AB, rises only through the vertical height CB, while P must move through a distance equal to AB. Let L represent

Fig. 59.

Fig. 60.

the length of an inclined plane, and H its height, and prepare a special formula for the solution of problems pertaining to the inclined plane.

18. A skid 12 feet long rests one end on a cart 3 feet high, and the other end on the ground. What force must a boy

exert while rolling a barrel of flour weighing 200 lbs. over the skid into the cart?

19. During one revolution a screw advances a distance equal to the distance between two turns of the thread, measured in the direction of the axis of the screw. Suppose the screw in the letter-press (Fig. 59) to advance ¼ inch at each revolution and a power of 25 lbs. to be applied to the circumference of the wheel b, whose diameter is 14 inches, what pressure would be exerted on articles placed beneath the screw? [The circumference of a circle is 3·1416 times its diameter.]

Fig. 61.

20. The toggle-joint (Fig. 60) is a machine employed where great pressure has to be exerted through a small space, as in punching and shearing iron, and in printing-presses, in pressing the types forcibly against the paper. An illustration may be found in the joints used to raise carriage-tops. Force applied to the joint c will cause the two links ac and bc to be straightened, or carried forward to d, while the guides move through a distance equal to $(ac + bc) - ab$. If $dc = 10^{cm}$, $ab = 98^{cm}$, and $ac + bc = 100^{cm}$, then a force of 80^g applied at c would exert what average pressure on obstacles in the path of the guides?

Fig. 62.

21. How would you calculate the mechanical advantage gained by a machine like that of Fig. 61? (On the axle A is

an endless screw, by means of which motion is communicated from the axle to the wheel W.)

22. (*a*) What kind of machine is a claw-hammer (Fig. 62)? (*b*) What mechanical advantage is gained by it?

23. In its technical meaning, a "perpetual motion machine"

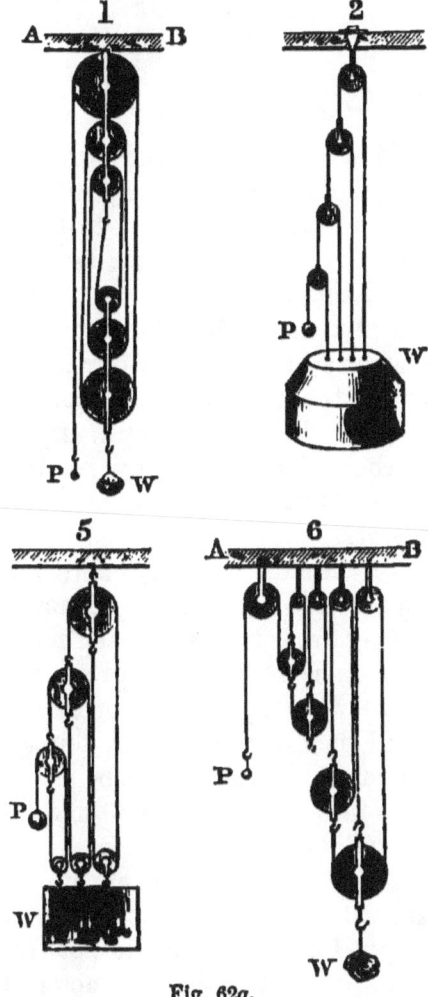

Fig. 62*a*.

is not a machine that will run indefinitely, but a machine which *can do work without having work done upon it*. Is such a machine possible?

24. A plank 12 feet long and weighing 24 lbs. is supported by two props, one 3 feet from one end, and the other 1 foot from the other end. What is the pressure on each prop?

25. With a single movable pulley what force will support a weight of 100 lbs.?

26. The gradient of a certain smooth road on a hillside is 1 foot in 10 feet. What force must a horse exert on a carriage which weighs together with its load one ton, to prevent its descent?

27. What must be the diameter of a wheel in order that a force of 20 lbs. applied at its circumference may be in

equilibrium with a resistance of 600 lbs. applied to the circumference of its axle, which is 3 inches in diameter?

28. Assuming the tension of the same string to be the same throughout its length and neglecting the weights of the pulleys, find the ratio of W to P in each of the combinations represented in Fig. 62*a*.

HYDROSTATICS

§ 88. Pressure in Fluids.—We have already learned (§ 16) that liquids and gases alike possess no rigidity of form. In consequence of this many laws apply to both. We shall, therefore, treat them together, in so far as they are alike, under the common name of *fluid*.

It should be borne in mind that we are placed on the borders of two oceans. A watery ocean borders our land; an aerial ocean, which is called the atmosphere, surrounds us. Every molecule, in both the gaseous and liquid oceans, is impelled toward the earth's centre by gravity. This gives to both fluids a downward pressure upon everything upon which they rest.

The gravitating power of liquids is everywhere apparent, as in the fall of drops of rain, the descent of mountain streams, the power of falling water to propel machinery, and the weight of water in a bucket. But to prove the downward pressure of air requires special experiments. If we lower a pail into a well, it fills with water, but we do not perceive that it becomes heavier thereby; the downward pressure is not felt. But when we raise a pailful out of the water, it suddenly becomes heavy. If we could raise a pailful of air out of the ocean of air, might not the weight of the air become perceptible? If we dive to the bottom of a pond of water, we do not feel the weight of the pond resting upon us. We do not feel the weight of the atmospheric ocean resting upon us; but we should remember that our situation with reference to the air is like that of a diver with reference to water.

Experiment 1.—Fill two glass jars (Fig. 63) with water, A having a glass bottom, B a bottom provided by tying a piece of sheet-rubber tightly over the rim. Invert both in a larger vessel of water, C. What force sustains the water in A? What can be the source of this force? Has the air anything to do with it? To satisfy yourself on this point, repeat the experiment, using mercury instead of water; place the whole apparatus under the receiver of an air-pump, and exhaust the air. What is the result, and how does it answer the question? The only reason for using mercury instead of water is that, on account of the greater density of the former, a visible result is obtained with less difficulty. Why is the india-rubber bottom of B forced inward?

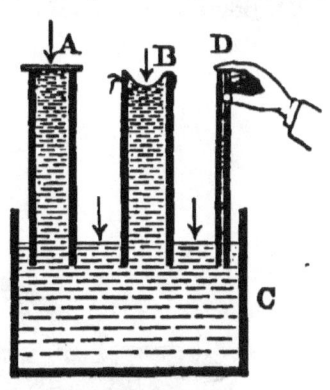

Fig. 63.

Take a glass tube, D, 1^m long, having a bore of 1^{cm} diameter. Covering one end with a finger, fill with water, and invert in C. What do you feel? Why? Remove the finger. What is the result? Why?

From the experiment with the mercury we learn that the downward pressure of the air on the surface of a liquid causes an upward pressure in the liquid. Does this downward pressure create an upward pressure in the air itself, so that, if the vessels are lifted out of the water, the water will not fall out?

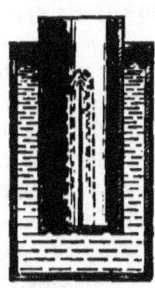

Fig. 64.

Experiment 2.—Keeping the finger pressed on the end of D, raise it slowly and vertically out of the water. Slip a thin glass plate, or a piece of thick pasteboard, under the mouth of A, and, pressing it against the mouth, raise the vessel carefully out of the water, and remove the hand from the plate. What answer do these experiments give to the question?

Experiment 3.—Force a tin pail (Fig. 64), having a hole in its bottom, as far as possible into water, without allowing water to enter at the top. What is the result? Why? Why does it require so much effort to force the pail down into the water?

Does downward pressure cause a lateral pressure?

Experiment 4.—Make holes, at different depths, in the

Fig. 65.

side of a vessel (Fig. 65) containing water. What does the experiment tell you?

Experiment 5.—Bind a piece of thin sheet-rubber tightly over a wide-mouthed bottle, and place it at considerable depth in water in different positions. Repeat the experiment at different depths. What is the result no matter in what position the bottle is placed? What facts do these experiments demonstrate?

Experiment 6.—The Madgeburg hemispheres (Fig. 66) are two hemispherical cups, having their edges made smooth so as to be "air-tight" when placed in contact. Each cup is provided with a handle. One of the handles consists of two parts, a stem and a ring, the two parts being connected by a screw. The stem has a bore passing through it, and a stop-cock, which regulates the passage of air through the bore. Place the lips of the cups in contact, first rubbing them slightly with tallow to ensure a perfect fit, remove the ring, screw the stem to the plate of an air-pump, and exhaust the air from the sphere; then close the stop-cock, and replace the ring. Let two boys grasp the rings, and, holding the sphere in any

Fig. 66.

position they choose, attempt to pull it apart. Are the hemispheres more easily separated when pulled in one direction than when pulled in any other?

Boys amuse themselves by lifting bricks (Fig. 67) with a circular piece of leather, moistened and pressed against the surface of the brick, so as to exclude the air.

Fig. 67.

Experiment 7.—Take a glass tube bent in the form represented by a, Fig. 68; place mercury in the lower part of the tube, so as to fill the short arm, and gradually lower the tube into a deep vessel of water. Sink the tube to different depths, and carefully watch the column of mercury. Prepare a two-column table. In one column set down the depths of the mercury surface in contact with the water. In the other column set down the differences between the heights of the two surfaces of the mercury. From a study of this table what general conclusion do you reach? Compare the phenomenon when the wide-mouthed bottle of Experiment 5 was sunk to different depths.

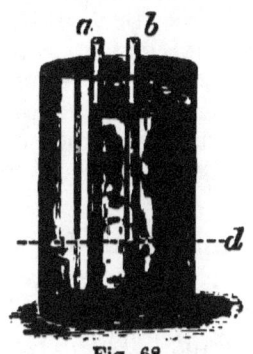

Fig. 68.

Experiment 8.—Introduce another tube, containing mercury, of the form represented by b, Fig. 68; lower both tubes so that the orifices in the water shall be at the same level. Observe the column of mercury in each tube. What does the experiment prove? How does your conclusion agree with that indicated by the phenomenon in Experiment 5, when the bottle, while kept at the same depth, was placed with its india-rubber covered mouth facing in different directions?

Experiment 9.—Cover one end of a lamp-chimney (Fig. 69) with a circular piece of leather, and suspend from the hand by means of a string attached to the centre of the leather and passing through the chimney. Hold the leather firmly against the bottom of the chimney, and lower the covered end a little way into a vessel of water. You may now drop the string, and the upward pressure of the water will keep the leather in place. Pour

Fig. 69.

water slowly into the chimney until the leather falls. What height does the water reach in the lamp-chimney before the leather falls? Why does it then fall? Why does not a pailful of water in a well seem heavy?

The results of experiments thus far show that, *at every point in a body of fluid, gravity causes pressure to be exerted equally in all directions, and that in liquids the pressure increases as the depth increases, the increase of pressure varying directly as the increase of depth.* It should also be observed that, on account of its mobility, *the direction of the pressure of a fluid against a surface in contact with it is at right angles to the surface.*

Experiment 10.—Place apparatus like Fig. 68 under the receiver of an air-pump and exhaust the air. Does the removal of the air cause any change in the position of the mercury in the tubes? If it does not, how is the fact to be explained? By removing the air we evidently remove pressure from the surface of the mercury in the long branch of the tube; but, if the mercury does not move, we must reduce *by just the same amount* the pressure of the water on the surface of the mercury in the other branch of the tube.

From this and other similar experiments we are led to the following very important conclusion: *If any pressure is communicated to the surface of a fluid, this pressure is transmitted equally to all points in the fluid.* This principle or law may be more clearly stated thus: *If the pressure*[1] *at any point in a fluid is increased (or diminished) by p pounds per square inch, the pressure at all points in that fluid is increased (or diminished) by p pounds per square inch.*

The foregoing law of fluid pressure may be seen to be a necessary consequence of the *mobility* and *perfect elasticity* of fluids. This may be illustrated, somewhat imperfectly, as follows:—

Fig. 70 represents a number of elastic hoops enclosed

[1] By the pressure at a given point in a fluid is meant the average pressure on an infinitely small area containing that point.

in the vessel ABCD. A weight, placed on a, communicates to it a downward pressure. It is evident that not only is the pressure communicated to the hoops below it in succession, and finally to the bottom of the box, but there is also a lateral pressure due to the elastic property of the hoops. The hoop c, receiving pressure from b, above, reacts, exerting an upward pressure; it also presses laterally upon the side A, and the hoop n, and downward upon d; d and n in turn transmit pressure to their adjacent hoops, and thus every hoop receives and transmits, upward, downward, and laterally, a force equal to the downward pressure of the weight W. Hence that portion of the bottom immediately under the weight receives no greater pressure from W than an equal area of any other part of the bottom, or than an equal area of either of the sides, A and B, or the top C. The following is a verification of the law itself:—

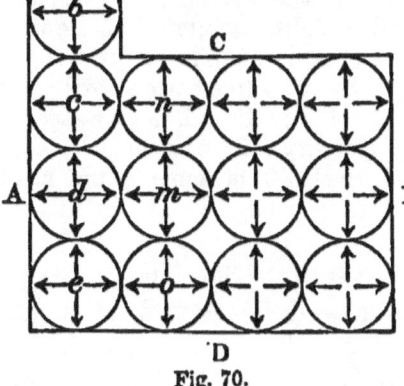

Fig. 70.

If we take a quantity of water in a vessel A (Fig. 71), shut in by two pistons, a and b, whose areas are respectively 16^{qcm} and 4^{qcm}, and place a 10-gramme weight on the platform d, and an equal weight on the platform c, it will be found that the latter is not sufficient to balance the former, but that it will require a 40-gramme weight placed on c to preserve equilibrium. But the area

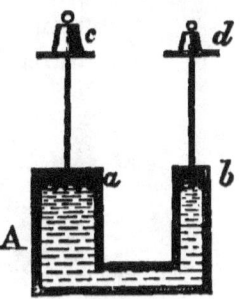

Fig. 71.

of the piston b is 4^{qcm}, while the piston a contains four such areas; hence it follows that a pressure of 10^g is transmitted to each of the 4^{qcm} of a, and just supports the 40-gramme weight. Had the area of the piston b been 1^{qcm}, then the 10-gramme weight placed on it would require a 160-gramme weight placed on a to balance it; that is, a pressure of 10^g would be exerted on every square centimetre of a.

Obviously this form of apparatus cannot be made to work well on account of the friction of the pistons; but we may substitute for the pistons and weights columns of liquids. For instance, let the connecting tube and the lower part of the barrels be filled with mercury; the two free surfaces will rest at the same level. Now if 10^g of any liquid, *e.g.* water, is poured into b, the level of the mercury will be changed; and, to bring it back to its original level, 40^g of some liquid must be poured into a.

§ 89. Hydrostatic Bellows.

The law of transmissibility of fluid pressure is well illustrated by means of the *hydrostatic bellows*. Two boards, b and c (Fig. 72), each having an area of (say) 400^{qcm}, are so connected, by leather attached to their edges, as to form an air-tight vessel called the bellows. A glass tube a, having a bore of 1^{qcm} section, communicates with the interior of the bellows. Let water be poured into the tube a till the board b is raised a few centimetres. The water will stand at almost the same height in the tube and bellows. Now, if 50^g of water be poured into the tube, it will require a weight of $20,000^g$ to be placed upon b to prevent its rising. Any weight less than that will be raised by the 50^g of water. If, instead of water being introduced into the bellows, a person stand on b, and blow into the

Fig. 72.

tube, he can easily raise himself by the pressure of the air from his lungs.

§ 90. **Hydrostatic Press.**—Closely allied to the bellows is the *hydrostatic press*, sometimes called *Bramah's press* from the name of the inventor. You see two pistons, t and s, Fig. 73. The area of the lower surface of t is (say) one hundred times that of the lower surface of s. As the piston s is raised and depressed, water is pumped up from the cistern A, forced into the cylinder x, and exerts an upward pressure against the piston t one hundred times greater than the downward pressure exerted upon s. Thus, if a pressure of one hundred pounds is applied at s, the cotton bales will be subjected to a pressure of five tons.

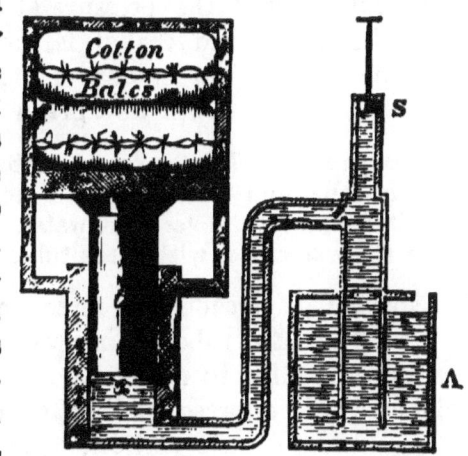

Fig. 73.

The pressure that may be exerted by these presses is enormous. The hand of a child can break a strong iron bar. But observe that, although the pressure exerted is very great, the upward movement of the piston t is very slow. In order that the piston t may rise 1^{cm}, the piston s must descend 100^{cm}. The disadvantage arising from slowness of operation is little thought of, however, when we consider the great advantage accruing from the fact that one man can produce as great a pressure with the press as a hundred men can exert without it.

The press is used for compressing cotton, hay, etc., into bales, and for extracting oil from seeds. The modern engineer finds it a most efficient machine, whenever great weights are to be moved through short distances, as in launching the *Great Eastern* steamship, or in raising into position the tubes of the Menai Bridge.

§ 91. Intensity of the Pressure at a given Point in a Liquid.—We have already learned that in a particular liquid the pressure at any point is the same in all directions, and that it is the same at all points at the same depth. Let us now try to learn what this pressure is at a given depth in a given liquid.

Experiment.—Take a *straight* tube, of *uniform bore*, open at both ends, and having a cross section of (say) 1^{qcm}, and place it *vertically* in a vessel of water. Observe the height to which the water rises within the tube.

Now let us consider the forces acting on the column of water within the tube. This water is at rest, and hence the forces tending to move it downwards are just counterbalanced by the forces tending to move it upwards. The sides of the tube are vertical, hence their pressure on the column of water within, being horizontal, has no effect to cause motion upwards or downwards. The only other forces acting upon this water are the pressure on its upper surface, its own weight, and the pressure on its base. It is evident therefore that the pressure of the surrounding water on its base is equal to the pressure on its upper surface plus its weight. Thus we learn from this experiment that *the pressure on one square centimetre at a point in water 1 centimetre below its surface is equal to the pressure on one square centimetre of the surface of this water plus the weight of 1 cubic centimetre of water*. Make similar experiments with other liquids, and state a general conclusion regarding the subject of this section.

§ 92. The Surface of a Liquid at Rest is Horizontal.—By jolting a vessel the surface of a liquid in it may be made to assume the form seen in Fig. 74. Can it retain this form? Take two particles of the liquid at the points a and b, on the same horizontal level. The downward pressure upon a is measured by $p + ac$, where p measures the pressure of the atmosphere on the surface of the liquid, and the downward pressure upon b is measured by $p + bd$. Now, since the pressure at a given depth is equal in all directions, $p + bd$ and $p + ac$ represent the lateral pressures at the points b and a respectively. But bd is greater than ac; hence the particles a and b, and those lying in a straight line between them, are acted upon, along this line, by two unequal forces in opposite directions, and, since ab is *horizontal*, the *weight* of these particles has no tendency to cause them to move along this line. There will, therefore, be a movement of liquid in the direction of the greater force toward a, till there is equilibrium of forces, which will occur only when the points a and b are equally distant from the surface; or, in other words, *there will be no rest till all points in the free surface are on the same horizontal level.*

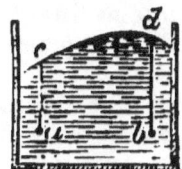

Fig. 74.

This fact is commonly expressed thus: "Water always seeks its lowest level." In accordance with this principle,

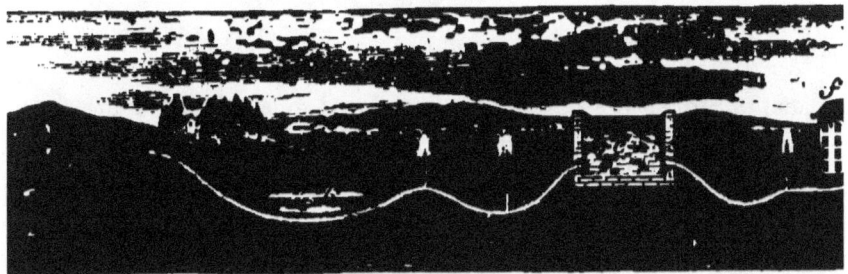

Fig. 75.

water flows down an inclined plane, and will not remain heaped up. An illustration of the application of this principle, on a large scale, is found in the method of supplying cities with water. Fig. 75 represents a modern aqueduct, through which water is conveyed from an elevated pond or river *a*, beneath a river *b*, over a hill *c*, through a valley *d*, to a reservoir *e*, in a city, from which water is distributed by service-pipes to the dwellings. The pipe is tapped at different points, and fountains rise theoretically to the level of the water in the pond, but practically not so high, on account of the resistance of the air and the check which the ascending stream receives from the falling drops. Where should the pipes be made stronger, on a hill or in a valley? Where will water issue from taps with greater force, in an upper chamber or in a basement? How high may water be drawn from the pipe in the house *f*?

§ 93. Artesian Wells, etc.—In most places the crust of the earth is composed of distinct layers of earth and rock of various kinds. These layers frequently assume concave

Fig. 76.

shapes, so as to resemble cups placed one within another. Fig. 76 represents a vertical section exposing a few of the surface-layers of the earth's crust; *a* is a stratum of loose sand or gravel; *b*, a clay-bed; *c*, a stratum of slate; *d*, a stratum of limestone; the whole resting on a bed of granite *e*. If you hollow out a lump of clay, and pour water into the cavity, you will find that the water will percolate through the clay very slowly. Water that falls in rain passes readily through the gravel *a*, till it reaches the clay-bed *b*, where it collects. Hence a *well*, sunk to the clay-bed, will fill with water as high as the water stands above the clay. Water also works its way from elevated places down between the strata of rocks. If a hole is bored through the slate c, water will rise above the surface of the ground in a fountain, in attempting to reach the level of its source on the hill; and if bored still lower, through the stratum d, a still higher fountain may result. Such borings are called *Artesian wells*. Water frequently forces its way through fissures in the rocky strata to the surface, as at t, and gives rise to *springs*.

QUESTIONS AND PROBLEMS

1. At high tide, suppose the flood-gate of a dock to be closed, leaving the surface of water on the inside and outside of the gate at the same level. From which does the gate sustain the greater pressure, the water in the dock, or the ocean of water outside? Why?

2. The interior dimensions of the rectangular vessel (Fig. 77) are 25^{cm} in length, 20^{cm} in width, and 15^{cm} in depth. The vessel is full of water. Compute the total pressure on its base.

3. Suppose that the plug n (Fig. 77), the area of whose end is 4^{qcm}, is pressed down upon the surface of the water with the force of 100^g; what additional pressure will each side of the vessel sustain?

Fig. 77.

4. Suppose mercury, which is 13·6 times as dense as water, to be employed instead of water, what would be the answers to the two preceding questions?

5. Into the top of a keg filled with water, a brass tube 10^m long is inserted, a transverse section of whose bore is 1^{qcm}. The depth of the water in the cask is 30^{cm}, and the area of the bottom of the cask is 40^{qcm}. (*a*) Compute the pressure on the bottom of the keg. (*b*) Compute the pressure on the bottom of the cask if the tube is filled with water. (*c*) What is the weight of the water in the tube that causes this extra pressure?

6. What crushing-force on each side would an empty cubical box, the area of one of whose sides is 1^{qm}, be subject to, if lowered 1^{km} into the sea?

7. What crushing-force on each side would this box be subject to from the atmospheric pressure at the sea-level, if the air were completely exhausted therefrom?

8. Suppose the top of the vessel (Fig. 77) to be the weak part of the vessel, not able to sustain more than 50ᵍ pressure on 10ᑫᶜᵐ, what pressure applied to the plug will burst the vessel?

§ 94. The Pressure of the Atmosphere.—

Experiment 3.—Prepare a U-shaped glass tube closed at one end (Fig. 78), 80ᶜᵐ in height from the centre of the bend, and with a bore of 1ᑫᶜᵐ section. Fill the closed arm with mercury, being careful to drive out *all* the air, and invert. The mercury falls to the level A. Carefully measure AB, while the tube is *perfectly vertical*. Now the surface at C is exposed to the pressure of the atmosphere, while the surface at A is exposed to no pressure. It is evident, therefore, that the pressure of the atmosphere on the one square centimetre of surface at C is equal to the weight of the mercury in the column AB. If AB is l centimetres long, the column of mercury AB, since its cross section is one square centimetre, has a volume of l cubic centimetres, and hence weighs 13,6 l grammes.

Fig. 78.

From your experiment what do you find to be the pressure of the atmosphere per square centimetre? From this conclusion determine, by calculation, its pressure per square inch and its pressure per square foot.

Would the column AB have a different length if a tube having a larger or a smaller bore were used?

Why not use water in making this experiment?

Why is it made a condition of this experiment that the tube shall be vertical when the length of the column AB is observed?

§ 95. The Barometer.—

Fig. 79 represents another form of apparatus, which is more commonly used for ascertaining atmospheric pressure. It consists of a straight tube about 85ᶜᵐ long, closed at one end, and filled with mercury. When this tube is inverted, the open end having been covered with a finger and plunged into an open cup of mercury, and the finger withdrawn, the mercury in the tube will sink till it balances the atmospheric pressure. This experiment was devised by Torricelli, an Italian. The apparatus is called a *barometer*.[1] The empty space above the mercury in the tube is called a *Torricellian vacuum*. The history of this experiment is very interesting and important, inasmuch as it was the first demonstration of the pressure of the atmosphere. (See Whewell's *History of Inductive Sciences*, vol. i. p. 345.)

The height of the barometric column is subject to fluctuations; this shows that the atmospheric pressure is subject to variations.

Fig. 79.

These variations arise from various causes. The barometer is always a faithful monitor of all changes in atmospheric *pressure*. It is also serviceable as a weather indicator. Not that any particular point at which mercury may stand foretells any particular kind of weather, but *any sudden change in the barometer indicates a change in the weather*. A rapid fall of mercury generally forebodes a storm, while a rising column indicates clearing weather.

If the barometer is carried up a mountain, it is found,

---

[1] Barometer, *weight measurer*.

150   ELEMENTS OF PHYSICS   CHAP.

as would be expected, that the mercury constantly falls as the ascent increases. This shows that the pressure

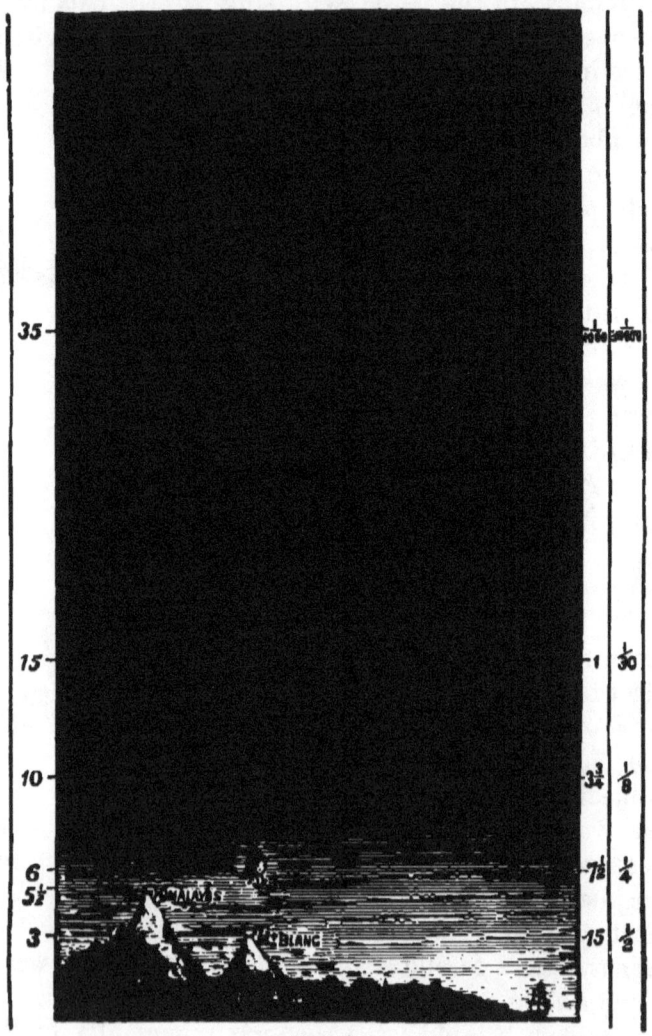

Fig. 80.

is greater near the bottom of the aerial ocean than near its top. It is found that the pressure increases very rapidly near the bottom, as may be understood by studying Fig. 80. The shading shows the variation in density of the air. The figures in the left margin show the height of the atmosphere, in miles; those on the right the corresponding height of the mercury, in inches. The average height of the mercurial column, at the level of the sea, is about 76^{cm} (30 inches).

It will be seen that the density at a height of 3 miles is but little more than $\frac{1}{2}$ the density at the sea-level; at 6 miles, $\frac{1}{4}$; at 9 miles, $\frac{1}{8}$; at 15 miles, $\frac{1}{30}$; at 35 miles it is calculated to be only $\frac{1}{30000}$, so that the greatest part of the atmosphere must be within that distance of the surface of the earth.

To what height the atmosphere extends is unknown. It is variously estimated at from 50 to 200 miles. If the aerial ocean were of uniform density, and of the same density that it is at the sea-level, its depth would be a little short of five miles. Certain peaks of the Himalayas would rise above it. It may be readily seen that heights of mountains may be measured approximately by the aid of a barometer.

A common form of barometer is represented in Fig. 81. Beside the tube and near its top is placed a scale graduated in inches or in centimetres indicating the height of the mercurial column.

§ 96. **Aneroid Barometer.**—In this instrument no liquid is used. It contains an air-tight cylindrical box D (Fig. 82), having a very flexible cover. The varying atmospheric pressure causes this flexible cover to rise and sink much like the chest of man in breathing. Slight movements

Fig. 81.

of this kind are communicated by means of multiplying-apparatus (apparatus by means of which a small movement of one part is magnified into a large movement of another part) to the index needle A. The dial is graduated to correspond with the scale of a mercurial

Fig. 82.

barometer. The observer turns the button C and brings the brass needle B over the black needle A, and at his next observation any departure of the latter from the former will show precisely the change which has occurred between the observations.

The aneroid can be made more sensitive (*i.e.* so as to show smaller changes of atmospheric pressure) than the mercurial barometer.

QUESTIONS

1. A person on the top of Mont Blanc would take in what portion of the air, on expanding his lungs to a certain extent, than he would at the bottom?

2. How would this affect breathing, considering that a person requires to inhale a definite weight or mass of air in a given time, in order to sustain life?

3. A person ascending 6 miles in a balloon leaves what fraction of the whole mass of air below him?

4. When the barometric column stands at 492^{mm}, what is the atmospheric pressure in grammes per square centimetre?

5. A barometer carried into a mine stands at 982^{mm}; what is the atmospheric pressure in the mine?

6. Would you expect a sudden change in the *pressure* of the atmosphere at a given point to be followed by a *movement of the atmosphere*, and hence by a *change of weather?* Why?

§ 97. Compressibility and Expansibility of Gases.

—The increase of pressure attending the increase in depth, in both liquids and gases, is readily explained by the fact that the lower layers of fluids sustain the weight of all the layers above. Consequently, if the body of fluid is of uniform density, as is very nearly the case in liquids, the pressure will increase in the same ratio as the depth increases. But the aerial ocean is far from being of uniform density, in consequence of the extreme compressibility of gaseous matter. For most practical purposes we may regard the density of water at all depths as uniform.

The pressure at different depths in liquids may be illustrated by piling several bricks one on another, when the pressures that different bricks sustain vary directly with their depths below the upper surface of the pile. On the other hand, pressure of gases at different depths may be illustrated by piling fleeces of wool one on another. Since the volume of each successive fleece varies with the weight it bears, the pressures which different fleeces sustain are not proportional to their respective depths below the upper surface of the pile. At twice the depth there is much more than twice the pressure, because the lower point supports more than twice the number of fleeces.

Closely allied to compressibility is the elasticity of gases, or their power to recover their former volume after compression. *The elasticity of all fluids is perfect.* By this is meant that the force exerted in expansion is always equal to the force used in compression; and that, however much a fluid is compressed, it will always completely regain its former bulk when the pressure is removed. Liquids are perfectly elastic; but, inasmuch as they are perceptibly compressed only under tremendous pressure, they are regarded as practically incompressible, and so it is rarely necessary to consider their elasticity. It has already been stated (§ 16) that matter in a gaseous state expands indefinitely, unless restrained by external force. What confines the atmosphere to the earth?

Experiment 1.—Partially fill an india-rubber balloon with air and tightly close it. What is the external force that prevents the air in the balloon from expanding and completely inflating the balloon? Place it under the glass receiver of an air-pump (Fig. 83) and exhaust the air. What is the result? Why?

Glass-blowers prepare thin glass bottles (Fig. 84) for the purpose of illustrating the tension of air. Containing air of ordinary density, they are sealed and placed under the receiver of an air-pump; the surrounding air (and hence the outside pressure) is removed, and the enclosed air then bursts the bottles, throwing fragments of glass in all directions.

Fig. 83.

Fig. 84.

Experiment 2.— Take a glass tube (Fig. 85), having a

bulb blown at one end. Nearly fill it with water, so that when inverted there will be only a bubble of air in the bulb. Insert the open end in a glass of water, place under an air-pump receiver, and exhaust. What happens? Why?

Fig. 85.

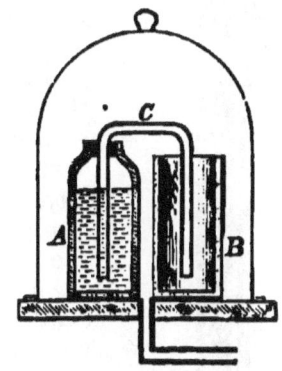

Fig. 86.

What will happen when the air is admitted to the receiver? Try it.

Experiment 3.—Through the cork of a tightly-stopped bottle pass one arm of a U-shaped glass tube C (Fig. 86). Introduce the other arm into the empty vessel B. Place the whole under a glass receiver, and exhaust the air. What phenomena will occur? What will happen when air is admitted to the receiver?

At every point, then, in a body of air, forces are acting outwards. The air is somewhat like a spring coiled up, and ready to relax itself, when opportunity is given. Since this elastic force at the bottom of the column exactly balances the force of gravity acting on the whole column, *i.e.* is equal to the weight of the whole column, it follows that, *at the sea-level, the elastic force of air is ordinarily 1^k per square centimetre* (§ 94).

§ 98. Boyle's and Mariotte's Law.—The foregoing experiments showed that the volume of a given body of gas depends upon the pressure to which it is subjected. To find more exactly the relation between these quantities, proceed as follows:—

156   ELEMENTS OF PHYSICS   CHAP.

Experiment 1.—Take a bent glass tube (Fig. 87), the short arm being closed, and the long arm, which should be at least 85^{cm} long, being open at the top. Setting it in a *vertical* position, pour mercury into the tube till the surfaces in the two arms stand at zero. Now the surface in the long arm supports the weight of an atmosphere. Therefore the tension of the air enclosed in the short arm, which exactly balances it, must be just that of the atmosphere outside. Next pour[1] mercury into the long arm till the surface in the short arm reaches 5, or till the volume of air enclosed is reduced *one-half*. Measure accurately the *vertical* distance between the surface of the mercury in the short arm and that in the long arm. How does this compare with the height of the barometric column at the time the experiment is made? What pressure does the air in the short arm now sustain? At first the air in the short arm sustained a pressure of one atmosphere. This air has been compressed to one-half its original volume. What relation do you find in this case between the volume of a given mass of air and the pressure to which it is subjected?

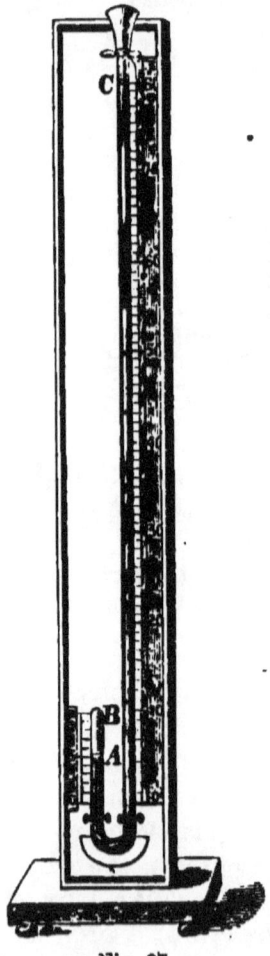

Fig. 87.

Experiment 2.—Next take a glass tube (Fig. 88) open at both ends, and about 24 inches long. Tie three strings around the tube,—one 3 inches from the top, another 6 inches, and the third 21 inches. Nearly fill a glass jar, B, 25 inches high with mercury. Lower the tube into the mercury till it reaches the string at 3. Press a finger firmly over the upper end, and raise the tube till the string at 21 is on a level with the surface of the mercury in the jar. Observe the space

---

[1] Pour this mercury in gently, so as not to drive any more air into the short arm.

now occupied by the air. Measure the height of the mercury in the small tube above the surface of the mercury in the large tube. What pressure does the air in the small tube now sustain? What relation do you find, in this case, between the volume of a given mass of air and the pressure to which it is subjected? Repeat the experiments varying the measurements.

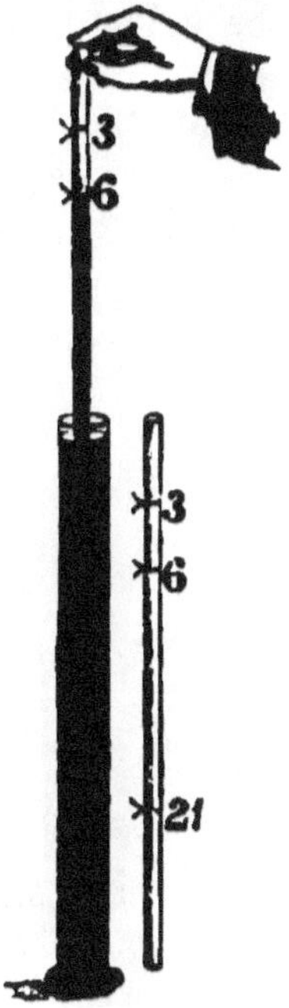

From experiments like the foregoing, with air and with other gases it has been found that, *at constant temperature, the volume of a body of gas varies inversely as the pressure.* This is sometimes called Mariotte's, and sometimes Boyle's, law, from the names of the two men who discovered it at about the same time. This law is true for all gases within certain limits, but under extreme pressure the reduction in volume is greater than asserted by it. The greatest deviation from it occurs with those gases that are most easily liquefied.

Having acquired precise knowledge regarding the compressibility and expansibility of gases, we are now in a position to understand the action of the air-pump, an instrument we have had occasion to use several times in our previous work.

Fig. 88.

§ 99. **The Air-pump.**—The air-pump, as its name implies, is used to withdraw air from a closed vessel. Fig. 89 will serve to illustrate its operation. R is a glass receiver from which air is to be exhausted. B is a hollow cylinder of brass, called the pump-barrel. A plug P, called a piston, is fitted to the interior of the barrel, and can be moved up and down by the handle H; *s* and *t* are valves. A valve acts on the principle of a door intended to open or close a passage. If you walk against a door on one side, it opens and allows you to pass; but if you walk against it on the other side, it closes the passage, and stops your progress. Suppose the piston to be in the act of descending. The compression of the air in B closes the valve *t*, and opens the valve *s*, and the enclosed air escapes. After the piston reaches the bottom of the barrel, it is drawn up; when the air above the piston, in attempting to rush down to fill the vacuum that is formed between the bottom of the barrel and the piston, closes

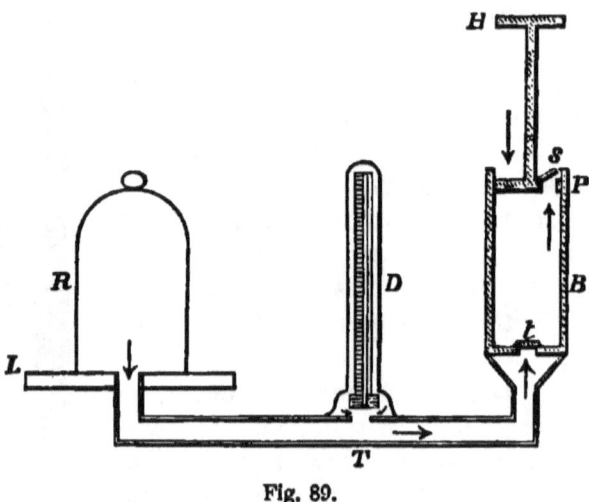

Fig. 89.

the valve *s*. But as soon as a vacuum is formed above *t*, and the downward pressure on the valve removed, the air in R expands, opens the valve *t*, and fills the space in B that would otherwise be a vacuum. But, as the air in R expands, it becomes rarefied; and, as there is less air, so there is less tension. The external pressure of the air on R, being no longer balanced by the tension of the air within, presses the receiver firmly upon the plate L. Each repetition of a double stroke of the piston removes a portion of the air remaining in R. The air is removed from R by its own expansion. However far the process of exhaustion may be carried, the receiver will always be filled with air, although it may be exceedingly rarefied. The operation of exhaustion is *practically* ended when the tension of the air in R becomes too feeble to lift the valve t.

D is another receiver, opening into the tube T, that connects the receiver with the barrel. Inside the receiver is placed a barometer. It is apparent that air is exhausted from D as well as from R ; and, as the pressure is removed from the surface of the mercury in the cup, the barometric column falls ; so that the barometer serves as a gauge to indicate the approximation to a vacuum. When the mercury has fallen 380^{mm} (15 inches), how much of the air has been removed ?

QUESTIONS

1. Why is it difficult for a person to lift the receiver from the pump after the air is exhausted from it ?
2. Why is it easily raised before the air is exhausted ?
3. Suppose that the air in the pump-barrel, when the piston is raised, is one-eighth of all the air in the pump, including the air in the receivers ; what portion of the air is removed by the first double stroke ?
4. What portion of the original amount of air is removed at the second double stroke ?
5. Which double stroke removes the most air ?
6. If there were no force required to lift the valve t, why could not a perfect vacuum be obtained?
7. It is a very good pump that reduces the height of the mercurial column to 3^{mm}. What portion of the air has been removed in that case ?
8. Into the neck of a bottle partly filled with water (Fig. 90) insert a cork very tightly, through which passes a glass tube nearly to the bottom of the bottle. Blow forcibly into the bottle. On removing the mouth, water will flow through the tube in a stream. Why ?

Fig. 90.

9. How can an ounce of air, in a closed fragile vessel, sustain the outside pressure of the atmosphere, amounting to several tons?

10. What drives the pellets from a pop-gun?

11. Fig. 91 represents a dropping-bottle, much used in chemical laboratories. Why do bubbles of air force their way down into the liquid?

12. Stop the upper orifice, and the liquid will quickly cease to drop. Why? The student can easily construct this apparatus for himself, making use of a short piece of large glass tubing, two pieces of small glass tubing, and two corks.

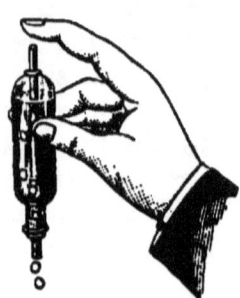

Fig. 91

13. The inconvenience arising, in many culinary and laboratory operations, from water "boiling away," may be remedied as represented in Fig. 92. A wide-mouthed bottle filled with water is so suspended that its mouth is just below the surface of the boiling liquid. As the water evaporates, and its surface falls below the mouth of the bottle, an air-bubble enters the bottle, expands, and pushes out enough water to cover once more the mouth of the bottle. Why does not the air push out all the water from the bottle?

14. Fig. 93 represents a weight-lifter. Into a hollow cylinder s is fitted air-tight a piston t. The cylinder is connected with an air-pump by an india-rubber tube u. When air is exhausted the piston rises, lifting the heavy weight attached to it. Why?

15. If the area of the lower surface of the piston is 20^{qcm}, how heavy a weight ought to be lifted when the air is one-half exhausted?

16. Suppose you were to tightly stopper a bottle at the top of Mont Blanc, carry it to the sea-level, insert the mouth of the bottle in water, and withdraw the stopper; what would happen?

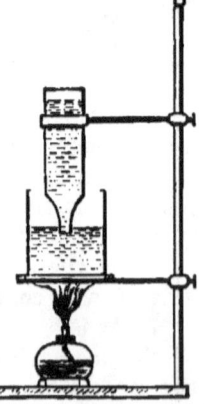

Fig. 92.

17. Show that the labour of working the kind of air-pump described (§ 99) increases as the exhaustion progresses.

An *absolute vacuum* has never been attained. The difficulty may be readily understood. According to the most recent calculations, the number of molecules contained in a cubic centimetre of air of ordinary density is something like 21,000,000,000,000,000,000; consequently, when it is reduced to one-millionth its usual density, 21,000,000,000,000 molecules are still left. The exhaustion may be carried much farther than by purely mechanical means, by heating a piece of charcoal in the receiver while the pumping is going on. Heat expels

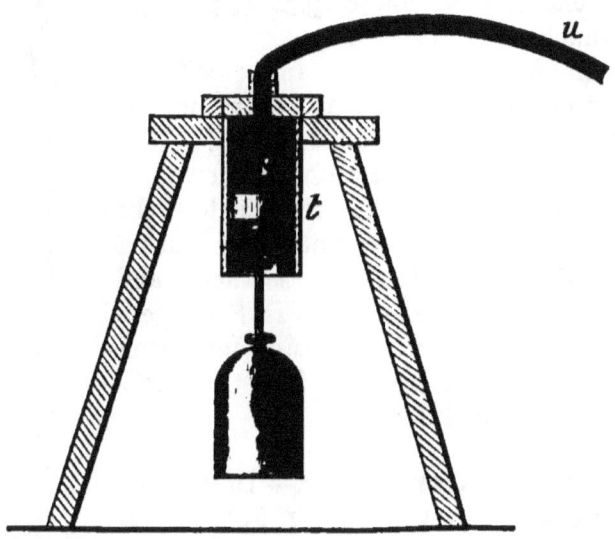

Fig. 93.

the air in its pores. After the pumping has ceased, the charcoal is allowed to cool, when it condenses a large portion of the remaining air in its pores (see § 37).

A very cheap and efficient substitute for an air-pump for many purposes may be arranged as in Fig. 94, in which a is an elevated tank of water having a tap b by which the rapidity of the flow of water may be regulated. The tube[1] c should be as long as the height of the room

[1] If the laboratory is connected with the town water-works the tube c may be connected with the service pipe. The writer has found such an arrangement very convenient for practical use.

will admit, and its lower end should dip into a cup of water *d*. To the end of the branch-pipe *e* there may be connected, by means of india-rubber tubing *h*, a glass tube leading to a vessel *g*, from which air is to be exhausted. Water falling freely through a vertical tube exerts no lateral pressure; consequently there is no tendency to enter the branch *e*. As the water in falling increases in velocity, it tends to separate, leaving between the cylinders of water vacuous spaces. The lower end of the pipe *c* being immersed in water, air cannot enter there; but the air in the receiver *g* expands and rushes through the tube *e*, to fill these vacua, and thus exhaustion is effected. In Sprengel's air-pump mercury is substituted for water, and air is reduced by it to less than one-millionth its usual density.

Fig. 94.

§ 100. **The Condenser.**—In the experiment with the bottle (Fig. 90), air was condensed in the mouth by muscular contraction, and forced into the bottle. An apparatus A (Fig. 95), intended to condense air in a closed vessel, is called a *condenser*. Its construction is like that of the barrel of the air-pump, except that the position of the valves is different. (Compare with Fig. 89.) What differences do you notice in respect to the valves? What happens to the valves when the piston in the condenser is forced down? If the condenser is connected with a closed vessel B, how much air is forced into it at one down stroke? What prevents the air from escaping during an up stroke? If, after air is condensed in B, the cylinder

C were connected with it by a screw, and the stop-cock *t* suddenly turned, what would happen to the bullet *s*? What name would you give to such an apparatus?

The Postal Telegraph Department, London, employs atmospheric pressure in forwarding messages to its central office from the various telegraph stations in that city. Tubes of uniform size, free from sudden curvatures, and laid under ground, connect the branch offices with headquarters. Rolls of paper, or letters to be despatched, are deposited in a cylindrical box *c* (Fig. 96), which fits the interior of the tube. The box being dropped into the end of the tube at *a*, and the air being exhausted from the tube at the end *b*, by means of an air-pump worked by steam, air rushes in at *a* and pushes the box through the tube with a force of several pounds for every square inch of the end of the box. The operation is still further facilitated by the aid of a condensing-pump worked by steam at the end *a*.

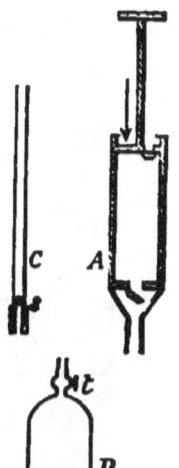

Fig. 95.

§ 101. **The Siphon.**—A siphon is an instrument used for transferring a liquid from one vessel to another through the agency of atmospheric pressure. It consists of a tube of any material (india-rubber is often most convenient for purposes of illustration), bent into a shape somewhat like an inverted U. To set it in operation, fill the tube with a liquid, stop each end with a finger or cork, insert one end in the liquid to be transferred, bring the other end below the level of the surface of this liquid, and remove the stoppers. What are the forces acting on the water in the siphon A (Fig. 97)? What forces tend to move the water from the tall jar to the short one? What forces tend to move the water the other way? By how much does the one set of forces exceed the other? When will the water cease to flow? Make a siphon by bending a glass tube, and, by means of two tall jars or bottles, experiment with it as in B and C. State and explain the results of your experiments.

The remaining diagrams in this cut represent some of the great variety of uses to which the siphon may be put. D, E, and F are different forms of siphon fountains. In D, the siphon tube is filled by blowing in the tube *f*. Explain the remainder of the operation. A siphon of the form G is always ready for use. It is only necessary to dip one end into the liquid to be transferred. Why does the liquid not flow out of this tube in its present con-

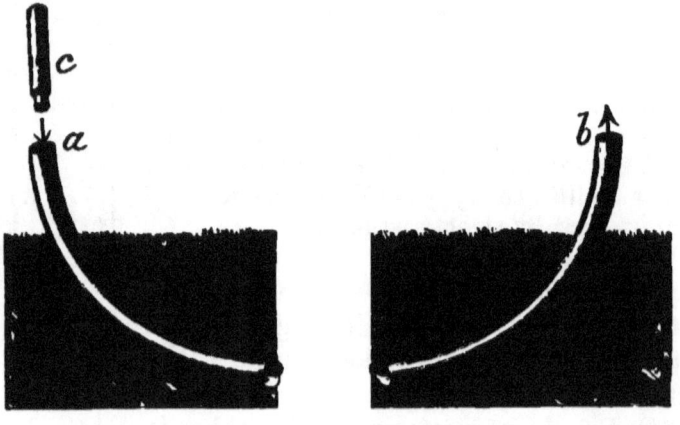

Fig. 96.

dition? H illustrates the method by which a dense liquid may be removed from beneath a less dense liquid. By means of a siphon a liquid may be removed from a vessel in a clear state, without disturbing sediment at the bottom. I is a *Tantalus cup*. A liquid will not flow from this cup till the top of the bend of the tube is covered. It will then continue to flow as long as the end of the tube is in the liquid. The siphon J may be filled with a liquid that is not safe or pleasant to handle, by placing the end *j* in the liquid, stopping the end *k*, and sucking the air out at the end *l* till the lower end is filled with the liquid.

Gases denser than air may be siphoned like liquids. Vessel *o* contains carbonic-acid gas. As the gas is siphoned into the vessel *p*, it extinguishes a candle-flame. Gases less dense than air are siphoned by inverting both the vessels and the siphon.

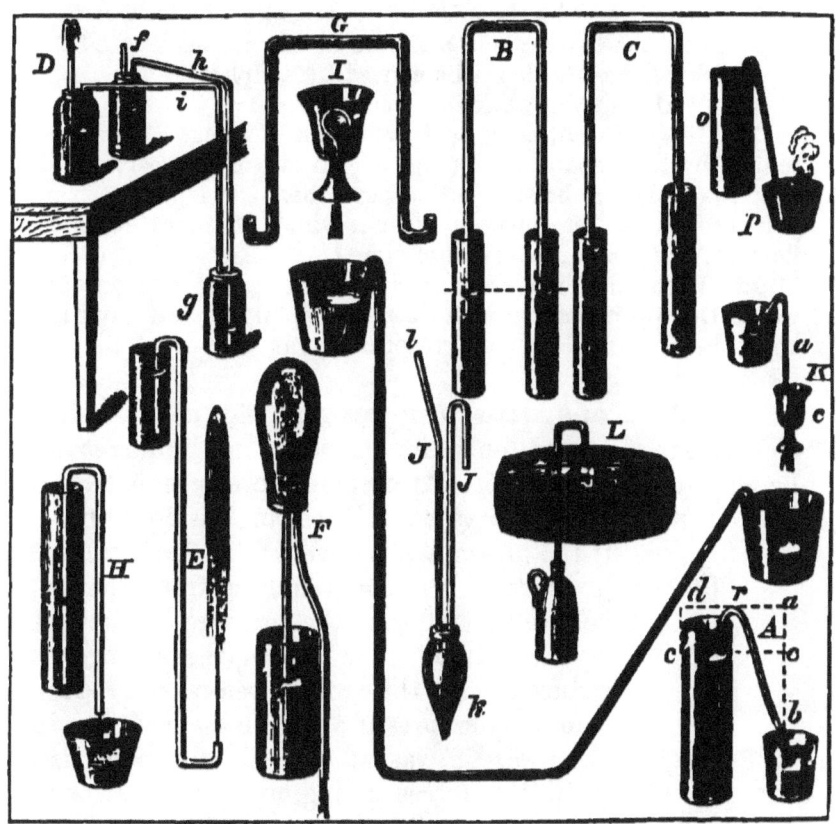

Fig. 97.

QUESTIONS

1. What is the greatest height to which the bend *r* (in A, Fig. 97) can be carried, and allow water to flow?
2. What would be the greatest height if mercury were used?
3. Suppose the bend *r* is 15^m above the surface of the liquid

in the tall jar; what, theoretically, ought to happen when the end *b* is unstopped?

4. What would happen if the long arm were cut off at *e*?
5. What would happen if it were cut off between *e* and *a*?
6. What would happen if the siphon were lifted out of the liquid?
7. What would be the effect of lengthening the long arm?
8. Must the two arms of a siphon be of unequal length?
9. How far can a liquid be carried by a siphon?
10. Will a siphon work in a vacuum?
11. Imagine that some such condition of things as is represented by the apparatus K (Fig. 97) exists in the earth, and that the siphon *a* has a smaller bore than the siphon *c*; can you account for intermittent springs which flow and cease to flow at nearly equal intervals of time?
12. If the siphon A were carried to the top of a mountain, would the water run through it more rapidly or less rapidly, or at the same rate? Your reason for your answer.

§ 102. Apparatus for raising Liquids.

—The siphon can be used only for transferring liquids over heights to a lower level. Atmosphere pressure is not a *store of energy* which may be utilised in raising liquids.

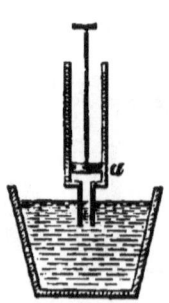

Fig. 98.

If the piston *a* of a *syringe* (Fig. 98) is raised, pressure is thereby removed from the air below it, and thus the pressure on the top of the column of water in the syringe is reduced. As this water, before the reduction in the pressure tending to force it downwards, was at rest, it will now rise until the forces acting upon it are once more in equilibrium; but to raise the piston, against the atmospheric pressure tending to force it downwards, requires as much muscular energy as would be required to raise the same quantity of water to the same height as that to which it is raised in the syringe.

The common *lifting-pump* is constructed like the barrel of an air-pump. Fig. 99 represents the piston in the act of rising. As the piston is raised, downward pressure is removed from the water below it, and this water rises and opens the lower valve in consequence of the pressure on its base due to the atmospheric pressure on the surface of the water outside the pump. The weight of the water above the piston closes the upper valve, and this water is discharged from the spout. When the piston is pressed down, the lower valve closes, the upper valve opens, and the water between the bottom of the barrel and the piston passes through the upper valve above the piston. How high may the bottom of the barrel be above the surface of the liquid, if the liquid to be pumped is water? How high if it is mercury?

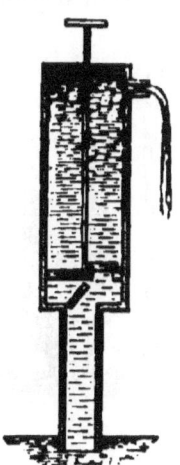

Fig. 99.

The liquid is sometimes said to be raised in a lifting-pump by the "force of suction." Is there such a force?

Experiment.—Bend a glass tube into a U shape, with unequal arms, as in Fig. 100. Fill the tube with a liquid to the level cb. Close the end b with a finger, and try to suck the liquid out of the tube. Remove the finger from b and try again. What is your conclusion?

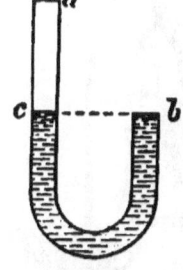

Fig. 100.

The piston of a *force-pump* (Fig. 101) has no valve, but a branch pipe leads from the lower part of the barrel to an air-condensing chamber a, at the bottom of which is a valve c, opening upwards. As the piston is raised, water is forced up through the valve d, while water in a is prevented from returning by the valve c. When the piston is forced down, the valve d closes, the valve c opens, and water is forced into the chamber a, condensing the air above the water. The elasticity of the condensed air forces the water out of the hose *b* in a continuous stream.

§ 103. **Buoyant Force of Fluids.**—**Experiment 1.**—Gradually lower a large stone, by a string tied to it, into a bucket of water, and notice the change in the muscular effort necessary to support it as it becomes submerged. Suspend the stone from a spring-balance, weigh it in air, and then in water, and ascertain its loss of weight in the latter. Repeat the experiment with pieces of iron, wood, and other substances. Inflate a bladder, and force it beneath a surface of water.

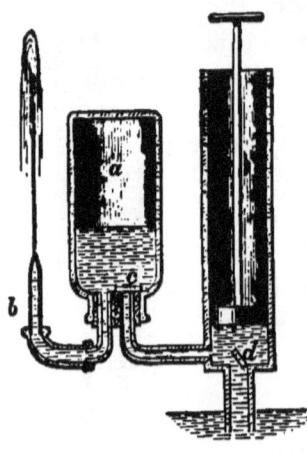

Fig. 101.

In all these experiments it seems as if something in the fluid, underneath the articles submerged, were pressing up against them. This lifting-force is called the *buoyant force* of fluids. Every body immersed appears to lose part of its weight; some bodies appear to lose all their weight. Is any weight lost?

Experiment 2.—Place a beaker of water on a scale-pan of a balance-beam, and weigh. Weigh a stone first in the air, then in water, and ascertain the apparent loss of weight. Then suspend the stone from a support (Fig. 102), and weigh the beaker of water with the stone immersed. How much more do the beaker and its contents now weigh than before? Compare this gain with the apparent loss of weight of the stone in water. Is any weight really lost? Repeat the experiment with a block of wood.

Fig. 102.

Experiment 3.—Make a saturated solution of salt in water. Weigh the same stone in air, in fresh water, and in salt water. Compare its apparent loss of weight in salt water with that in fresh water. Can you account for the difference? Throw a piece of iron into mercury. Fill a vessel with carbonic-acid gas; blow a soap bubble, and drop it into the vessel. It appears that some fluids have greater buoyant force than others. The water of the Dead Sea, in Palestine, is so salt that a person does not sink in it. Can you see the reason?

By considering what we have already learned about fluid pressure, it is easy to see why a body is buoyed up by a fluid. We have learned that the pressure increases with the depth, and hence can see that the upward pressure of the surrounding fluid on a submerged body must be greater than the downward pressure. Let us ascertain the *resultant pressure* of a fluid upon a body submerged in it.

Experiment 4.—Fill the vessel A (Fig. 103) till the liquid overflows at E. After the overflow ceases, place a vessel C

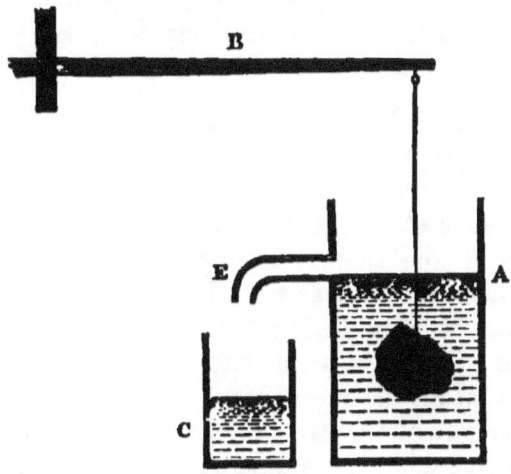

Fig. 103.

under the nozzle. Suspend a stone from the balance-beam B, and weigh it in air, and then carefully lower it into the liquid, when some of the liquid will flow into the vessel C. Weigh the liquid in C, and compare its weight with the apparent loss of weight of the stone when weighed in this liquid.

Experiment 5.—Next suspend a block of wood that will float in the liquid, and weigh it in air. Then float it upon the liquid, and weigh the liquid displaced as before, and compare this weight with the weight of the block of wood.

Experiment 6.—Next, partially fill with water a glass (Fig. 104) graduated in cubic centimetres and fractions of the same. Note the level of the water. Drop one of the solids into the water, and note again the level of the water. The difference between the two levels is the number of cubic centimetres of water that the solid displaces. But one cubic centimetre of water weighs one gramme. How does the number of cubic centimetres of water displaced compare with the number of grammes the solid apparently loses in weight when weighed in water?

Fig. 104.

From your observations in connection with these experiments prepare a statement of the law regarding the *buoyant force of fluids*.

It has been stated that the density of the atmosphere is greatest at the surface of the earth. A body free to move cannot remain at rest while it displaces more than its own weight of a fluid; therefore a balloon, which is a large bag filled with a gas having about one-fourteenth the density of air at the sea-level, will rise till the balloon, plus the weight of the car and cargo, equals the weight of the air displaced. The aeronaut, wishing to ascend still higher, throws out a portion of his cargo; wishing to descend, he allows some of the gas to escape at the top of the balloon by means of a valve, which he controls by means of a cord passing through the balloon to the car.

QUESTIONS

1. Why is it difficult to stand in running water reaching the neck?
2. Why can a person raise a stone under water, which he cannot lift when out of water?
3. A piece of cork weighs 50^g; what weight of water does it displace when floating?
4. What weight of mercury will a piece of iron weighing 500^g displace when floating?

§ 104. **Density.**—We speak of a piece of cork as being heavier than a nail, at the same time that we speak of cork as light and iron as heavy. This seeming contradiction is accounted for by the different meanings which we attach to the terms *light* and *heavy*. In both cases, *light* and *heavy* are used as terms of comparison. In the former instance, we compare the weights of the two particular bodies, without reference to volume; in the latter, we call cork *light* and iron *heavy*, having no particular bodies in view, but because we know by experience that cork is not so *dense* as iron, *i.e.* a given volume of cork weighs less than an equal volume of iron. The term *weight* refers simply to the number of grammes, kilogrammes, etc., that a particular body weighs, without reference to the material or the volume. The *density* of a body can be stated only by expressing (or understanding) two quantities, viz. mass and volume. For example, suppose that a block of wood measures $2 \times 10 \times 20^{cm}$ and has a mass of (*i.e.* weighs) 300^g; its density is then $\frac{300}{2 \times 10 \times 20} = \frac{300}{400} = 0.75$ gramme per cubic centimetre. When we speak of cork as lighter than iron, it is evident that we are comparing the *densities* of these two substances.

§ 105. **Specific Gravity.**—*The specific gravity of a substance is the ratio of the density of that substance to the density of another substance assumed as a standard;* in other words, *it is the ratio of the weight of any volume of that* *substance to the weight of the same volume of another substance which is taken as a standard.*

To facilitate comparison of densities, uniform standards are adopted. Distilled water at its maximum density, at 4° C., is the standard of specific gravity for all solids and liquids. Inasmuch as one cubic centimetre of water weighs one gramme, when the weight of one cubic centimetre of any substance is given in grammes, *i.e.* when its density is given in its usual metric units, the same number also expresses its specific gravity. Thus, one cubic centimetre of water weighs one gramme, and 1 is the specific gravity of water. The density of silver is 10.53^g per cubic centimetre, and the specific gravity of silver is 10.53. The standard for gases is air [1] at the average sea-level density, and at a temperature of 0° C. The weight of one cubic centimetre of air, under these conditions, is 0.0012932^g, or about $\frac{1}{773}$ of the weight of one cubic centimetre of water.

Let G = the specific gravity of a substance; D = its density in grammes per cubic centimetre; V = the volume of a given mass of it in cubic centimetres; W = the weight of the given mass in grammes; W' = the weight in grammes of an equal volume of the standard (water). Then, as shown above, $D = \frac{W}{V}$, and, by definition, $G = \frac{W}{W'}$. G is *numerically* equal to D, and W' to V.

Since the loss of weight of a solid immersed in a liquid is just the weight of an equal volume of that liquid, it is evident that, *if we divide the weight of a solid in air by its loss in weight when immersed in water, the quotient is its specific gravity.*

Experiment 1.—Obtain small lumps of glass, iron, lead, marble, granite, etc., and weigh each in air. Partly-fill with water a measuring-beaker graduated in cubic centimetres, and note the level of the water. Drop a lump into the water, and

---

[1] Some chemists adopt hydrogen gas as a standard for gases.

note the level again. The rise of water, as indicated by the graduated scale, gives the volume (V) of the specimen. With these data find the density (D), employing the formula $D = \frac{W}{V}$. Next weigh each of these lumps submerged in water, and find its loss in weight; and, from the data obtained, ascertain G from the formula $G = \frac{W}{W'}$. Prepare blanks, and tabulate your results thus:—

Name of Substance.	W g	V ccm	D or G	e	W g	W in water. g	W' g	G or D	e	Av.	e
Flint glass	435	134	3·24	·09 −	435	305	130	3·34	·01 +	3·29	·04 −

When the result obtained differs from that given in the table of specific gravities the difference is recorded in the column of errors (e). When the former is greater than the latter, it is indicated by a plus sign affixed to the number; when less, by the minus sign. The results recorded in the column of errors are not necessarily *real* errors; they may indicate the degree of impurity, or some peculiar physical condition, of the specimen tested.

Experiment 2.—Obtain good specimens of cork, oak, elm, and poplar woods, all of which float on water. Tie to a specimen a piece of lead heavy enough to sink it. Weigh the two in air and in water. The difference is the weight of water displaced by them. In the same way find the weight of water displaced by the lead alone. Subtract the weight of water displaced by the lead from the weight of water displaced by the two, and the remainder is the weight of water displaced by the specimen. Then apply the formula $G = \dfrac{W}{W'}$.

Example.—Find the specific gravity of a piece of elm wood. Attach to it a piece of lead weighing (say) 40^g.

The combined solids displace . . .	$28 \cdot 5^g$ of water
The lead displaces	$3 \cdot 5^g$,,
The elm displaces	$25 \cdot 0^g$,,
The elm weighs in air	$20 \cdot 0^g$,,
The specific gravity of elm wood is .	$20 \cdot 0 \div 25 = \cdot 8$

Experiment 3.—Find the specific gravity of alcohol, a saturated solution of common salt, sea-water, naphtha, olive-oil, pure milk, and mercury in the following manner: ascertain the loss in weight of a sinker in each one of these liquids, also in water, and then apply the formula $G = \dfrac{W}{W'}$. Here W and W' represent the loss of weight of the sinker in the liquid and in water respectively.

Example.—Compute the specific gravity of alcohol from the following data:—

A piece of marble weighs in air . .	$56 \cdot 80^g$
The same weighs in water . . .	$36 \cdot 80^g$
Loss in water	$20 \cdot 00^g$
	$56 \cdot 80^g$
The marble weighs in alcohol . . .	$40 \cdot 96^g$
Loss in alcohol	$15 \cdot 84^g$

Since 20^g and $15 \cdot 84^g$ are the weights respectively of equal volumes of water and alcohol, and since $G = \dfrac{W}{W'}$, then $\dfrac{15 \cdot 84}{20}$ $= \cdot 792$, the specific gravity of alcohol.

§ 106. Hydrometers.—Experiment.—Take a uniform rod of light wood about a foot long, and mark off on it a scale of equal parts. A convenient size is $\frac{1}{4}$ inch square, and a suitable scale is inches and half inches. Coat the rod with paraffin to prevent its absorbing water and swelling. Bore into the end marked zero a hole about 2 inches deep, and drive in leaden shot till the rod will sink in water (Fig. 105) just to some inch-mark, and stop the end with paraffin. If it sinks too deep, cut off the upper end of the rod.

Suppose the rod sinks 8 inches in water; then, if it is $\frac{1}{2}$ inch square, it displaces 2 cu. in. of water. The weight of the water displaced must just equal the weight of the rod (see § 103). Now immerse it in alcohol; it sinks deeper, say to the 10-inch mark; that is, $\frac{10}{4}$ cu. in. of alcohol weigh the same as $\frac{8}{4}$ cu. in. of water; therefore, $G = \frac{V}{V} = \frac{8}{10} = \cdot 800$. If in brine it sinks only $6\frac{2}{3}$ in., $G = \frac{8}{6\frac{2}{3}} = 1\cdot 20$.

Apparatus like that described is called a *hydrometer*. Instead of a rod of wood, a glass tube is generally used, terminating in a bulb containing shot or mercury. The tube contains a scale with numbers corresponding, which express the specific gravity, so that no computation is necessary. Make solutions of various substances, and test their specific gravity with your hydrometer, and test the accuracy of the results so obtained by other processes.

Fig. 105.

A hydrometer, called Nicholson's, can be readily made by any tinman. It consists (Fig. 106) of a cylindrical float, to the top of this is attached a stout wire that supports a plate upon which the weights and the body to be weighed can be placed. At the other extremity of the float is attached a little scale-pan in which the body can be immersed in the liquid. The float is weighted at d, so that it floats upright.

The body is first placed upon the plate with additional weights in order to sink the hydrometer to a fixed mark c upon the stem. The body is then removed, and the increase of weight necessary to again sink the hydrometer to the same mark will give the weight of the body in air.

The latter is then placed in the lower scale-pan. The weights in the upper plate, or pan, necessary to sink the hydrometer to the point *c*, will be greater than when the body was placed in the upper pan, by the weight of this liquid displaced by the body. It will thus be seen that this instrument may be used to find the specific gravity of solids.

How would you use it to find the specific gravity of liquids?

The most direct way of finding the specific gravity of liquids and gases is by employing vessels that hold definite weights of the two standards, water or air, and then weighing these vessels when filled with other liquids or gases; and, after deducting the weight of the vessel, applying the formula, $G = \dfrac{W}{W'}$.

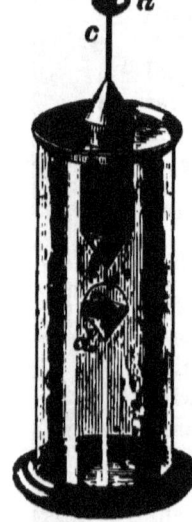

Fig. 106.

The specific gravity of a solid that is dissolved by water may be found by weighing it in a liquid that will not dissolve it (*e.g.* rock-salt in naphtha); and, having found its specific gravity as compared with the liquid used, multiply this result by the specific gravity of the liquid.

From the formula $D = \dfrac{W}{V}$, we have $V = \dfrac{W}{D}$; hence, *the volume of an irregular-shaped body may be found in cubic centimetres by dividing its weight in grammes by its density.*

Again, from the formula $D = \dfrac{W}{V}$, we have $W = V \times D$. Hence, *when the volume and density of a body are known, its weight in grammes may be found by multiplying its volume in cubic centimetres by its density.*

QUESTIONS AND PROBLEMS

1. How high can sulphuric acid be raised by suction?
2. What is the weight of 50^g of water in water?
3. Find the specific gravity of wax from the following data: weight of a given mass of wax in air is 80^g; wax and sinker displace $102\cdot 88^{ccm}$ of water; sinker alone displaces 14^{ccm}.
4. Why does a light liquid (*e.g.* oil), introduced under a denser liquid (*e.g.* water), rise?
5. Glass is about three times denser than water; how, then, can a glass tumbler float in water?
6. How can iron vessels float in water?
7. A block of ice containing 500^{ccm} is floating on water; how many cubic centimetres are out of water?
8. Will ice float or sink in alcohol?
9. How much more matter is there in 500^{ccm} of sea-water than in the same volume of fresh water?
10. In 50^k of gold how many cubic centimetres?
11. What is the density of gold?
12. What is the density of cork?
13. What is the density of air at ordinary pressure, and at a temperature of $0°$ C.?
14. An irregular piece of marble loses 53^g when weighed in water. How many cubic centimetres does it contain?
15. When will a body sink, and when float?
16. How many cubic centimetres of air at the sea-level does it take to weigh as much as 1^{ccm} of water?
17. How much will 1^k of copper weigh in water?
18. What does a piece of lead $20 \times 10 \times 5^{cm}$ weigh?
19. What will it weigh in water?
20. What will it weigh in mercury?
21. What becomes of the weight that is lost?
22. If 15^g of salt be dissolved in 1^l of water, without increasing the volume of the liquid, what will be the specific gravity of the solution?
23. A mass of lead weighs 1^k in air. What will it weigh in a vacuum?
24. A mass whose weight in air is 30^g, weighs in water 26^g, and in another liquid 27^g. What is the specific gravity of the other liquid?
25. A silver spoon, weighing 150^g, is supported by a string

in water. What part of the weight is sustained by the string, and what part is supported by the water?

26. A boat displaces 25^{cbm} of water. How much does it weigh?

27. If 50^k of stone were placed in the boat, how much water would it displace?

28. If the boat is capable of displacing 100^{cbm} of water, what weight must be placed in it to sink it?

29. An empty glass globe weighs 100^g; full of air it weighs $102\cdot4^g$; full of chlorine gas, it weighs $105\cdot928^g$. What is the specific gravity of chlorine gas?

30. What weight of alcohol can be put into a vessel whose capacity is 1^l?

31. You wish to measure out 50^g of sulphuric acid. To what number on a beaker graduated in cubic centimetres will that correspond?

32. State how you would measure out 80^g of nitric acid in a measuring-beaker.

33. A measuring-beaker contains 35^{ccm} of naphtha. What is the weight of the naphtha?

34. A lead pipe is used to convey water 20^m below the surface of the reservoir. What bursting-force per square centimetre must it be capable of sustaining?

35. A cubical vessel, each of whose sides contains 2500^{qcm}, is filled with water. What pressure does its bottom sustain?

36. A solid floats partly submerged in a liquid when the vessel which contains it is in the air; if the vessel is placed in a vacuum, will the solid sink, rise, or remain stationary? Why?

CHAPTER IV

HEAT

§ 107. Introductory Experiments.—Experiment 1.
—Heat a brass or an iron ball to a high temperature in the flame of a spirit lamp or in a gas flame, and drop it into a beaker of cold water. After a minute take it out. What change has taken place in the temperature of the ball in consequence of its contact with the water? What change has taken place in the temperature of the water? Repeat the experiment, beginning with hot water and a cold ball, and observe the changes of temperature arising from contact between the water and the ball.

State a general conclusion which you draw from this and other similar experiments.

Experiment 2.—Drive two nails firmly into a board about six inches apart, and cut off a piece of stiff copper wire of such a length that it will just pass between the two nails. Heat this wire in the flame of your spirit lamp and try to pass it between the two nails. What change do you find in the length of the wire? Cool the wire and try again. What do you now find?

Fig. 107.

Experiment 3.—Fit a stopper tightly in the neck of a glass flask, and through the stopper pass a glass tube a few inches long. Place this apparatus in a jar of water, as represented in Fig. 107, and apply heat to the flask. What takes place? Allow the flask to cool, and observe the result. What does the experiment prove?

Experiment 4.—Fill the same flask with water, and insert the stopper and the tube so that the water will stand in the tube an inch or two above the neck of the flask. Apply heat to the flask, and watch the water in the tube. What is the result? Cool the flask, and what do you see?

What general conclusion do you draw from Experiments 2, 3, and 4?

Experiment 5.—Carefully weigh the metal ball of Experiment 1, and, after heating it, weigh it again. How does its weight when hot compare with its weight when cold?

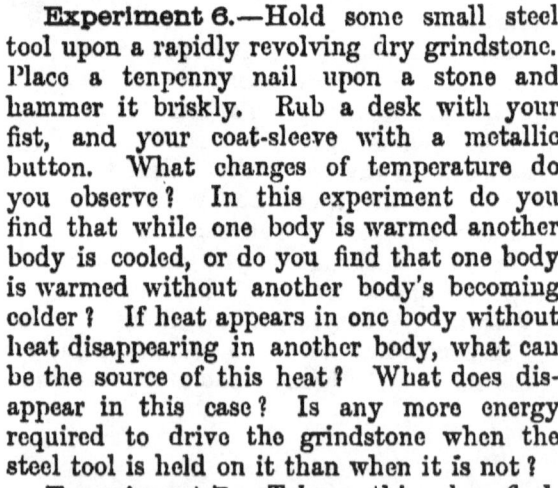

Fig. 108.

Experiment 6.—Hold some small steel tool upon a rapidly revolving dry grindstone. Place a tenpenny nail upon a stone and hammer it briskly. Rub a desk with your fist, and your coat-sleeve with a metallic button. What changes of temperature do you observe? In this experiment do you find that while one body is warmed another body is cooled, or do you find that one body is warmed without another body's becoming colder? If heat appears in one body without heat disappearing in another body, what can be the source of this heat? What does disappear in this case? Is any more energy required to drive the grindstone when the steel tool is held on it than when it is not?

Experiment 7.—Take a thin glass flask A (Fig. 108) and half fill it with water; fit a cork air-tight[1] in its neck. Perforate the cork, insert a glass tube, bent as indicated in the figure, and extend it into the water. Apply heat to the flask. What is the result?

In this experiment *work is done* by means of heat.

[1] A good way to make a cork air-tight is to soak it in melted paraffine.

The water is lifted against the force of gravity. Do you observe any connection between this and the preceding experiment? Here visible motion is produced from heat. How was it in Experiment 6? We cannot tell with this apparatus whether, when the heat does work, any heat disappears or not. To answer this question make the next experiment.

Experiment 8.—By means of a condenser (§ 100) force air into a receiver and close the stop-cock. Set the receiver aside until it and the contained air reach the *same temperature as the air of the room*. Now open the stop-cock, and allow the air as it rushes out to strike against the glass flask of the apparatus of Experiment 3. Is there any indication that this air has now a temperature different from that of the air of the room? As this air rushes out, *work is done* in pushing back the air of the room. Motion is produced against resistance (against the pressure of the surrounding air). What can be the energy expended in doing this work? What disappears as this work is done? Does the experiment show that any *heat* disappears?

Experiment 9.—Half fill a small glass beaker with fragments of ice or with snow, and set the beaker into a basin of boiling hot water, in which a thermometer is also placed. Stir the contents of the beaker with another thermometer until the ice is all melted. Carefully observe during this experiment the temperature of the contents of the beaker, and also the temperature of the water in the basin. Do you find that *while the ice is melting* any change takes place in the temperature of the contents of the beaker? Does any change take place in the temperature of the contents of the beaker *after the ice is all melted*? Does any change take place in the temperature of the water in the basin (*a*) *while the ice is melting*; (*b*) *after the ice is all melted*? Do you infer from this experiment that *heat is required to change ice into water*?

Experiment 10.—Take a glass test-tube half full of cold water, and pour into it one-fourth its volume of sulphuric acid. What is the effect? How does the volume of the mixture compare with the sum of the volumes of the acid and the water before they are mixed?

The following celebrated experiment was first made

by Sir Humphry Davy, and is described by him in his *Essay on Heat, Light, and Combinations of Light* :—

" I procured two parallelopipedons of ice, of the temperature of 29°, 6 inches long, 2 wide, and ⅔ of an inch thick ; they were fastened by wires to two bars of iron. By a peculiar mechanism their surfaces were placed in contact and kept in a continued and violent friction for some minutes. They were almost entirely converted into water, which water was collected, and its temperature ascertained to be 35° after remaining in an atmosphere of a lower temperature for some minutes. The fusion took place only at the place of contact of the two pieces of ice, and no bodies were in friction but ice."

In the study of this experiment the following conditions should be noted :—

1. During the experiment the bodies in contact with the ice were at a temperature *lower than that at which ice melts.*
2. The temperature of surrounding bodies *was not lowered* during the experiment.
3. Not only was most of the ice melted but the resulting water was warmed to 35°, notwithstanding that the surrounding atmosphere was at a lower temperature.
4. No bodies were in friction but ice.
5. Mechanical energy was expended (work was done) in rubbing the pieces of ice together.

Considering this experiment in connection with Experiments 1 and 9, what answers do you give to the following questions ?—

1. Did the ice in Davy's experiment receive any heat *taken from surrounding bodies* ?
2. Did it receive any heat from any source ?
3. If it did, what must be that source ?
4. Is your answer to Question 3 confirmed by the results observed in Experiments 6, 7, and 8 ?

§ 108. **Hypotheses as to the Nature of Heat.** —From the dawn of science till the close of the last century two rival hypotheses had been entertained regarding the nature of heat. Each accounted in a fairly satisfactory manner for the various phenomena of heat so far as these phenomena had then been observed, but neither rested on any sure experimental basis. One hypothesis supposed heat to be a subtle weightless fluid, called *caloric*, which permeated the spaces among the particles of matter, like water in a sponge. The caloric was supposed to flow from a hot to a cold body when the two are in contact. Its presence was supposed to increase the size of a body, hence expansion by heat. Friction or hammering was supposed to diminish the capacity of a body for caloric, and hence to cause the caloric to flow out. Combustion was supposed to send out heat because the capacity for caloric of the substance resulting from the combustion is less than that of the substances burning. Thus more or less plausible explanations were offered of all the then observed phenomena. But this hypothesis would not permit one to suppose *heat to appear in one body which did not previously exist as heat in that or in some other body*. It would be absurd to suppose this caloric to be *created* in any physical or chemical process, and this hypothesis did not suppose caloric and any other substance to be convertible one with the other. The other hypothesis supposed heat to be a commotion among the particles of matter. According to this hypothesis heat is *not matter but a form of energy*, the kinetic energy (§ 66) due to the motion of the particles or molecules of a body among one another. In the year 1799 Davy published the work in which his experiment with the ice is described, and thereby conclusively overthrew the former of these hypotheses, at the same time giving good reason for accepting as true the latter. Subsequent investigations have fully confirmed this conclusion, so that now we may with every confidence accept what is generally called *the dynamical theory of heat*.

Applying what you have learned from your study of dynamics, particularly regarding energy and work, show

how each of the various phenomena observed in the introductory experiments of this chapter is to be explained by the dynamical theory of heat. Which of these phenomena are not satisfactorily explained by the fluid hypothesis ?

Having made a brief qualitative study of heat, let us consider *temperature*, and means by which it may be accurately determined, that we may be enabled to proceed with more precise investigations.

§ 109. **Temperature defined.**—If body A is placed in contact with body B, and A loses and B gains heat, then A is said to have had originally a higher *temperature* than B. If neither body gains or loses then both had the same temperature. *Temperature is the state of a body considered with reference to its power of communicating heat to or receiving heat from other bodies.* The direction of the flow of heat determines which of two bodies has the higher temperature.

It may be mathematically demonstrated, using the laws of motion as a basis, that, if the average kinetic energy of each particle of A is equal to the average kinetic energy of each particle of B, then when A and B are brought into contact the particles of A will receive from the particles of B just as much energy as those of B receive from those of A. Hence, adopting the dynamical theory of heat, we may say that *two bodies have the same temperature when the average kinetic energy of each particle of the one is equal to the average kinetic energy of each particle of the other.*

§ 110. **Temperature distinguished from Quantity of Heat.**—The term *temperature* has no reference to *quantity* of heat. If we mix together two equal quantities of the same substance at the same temperature, the temperature of the mixture is not greater or less than that of either before they were mixed ; but evidently the mixture contains twice as much heat as either alone. If we dip from a gallon of boiling water a cupful, the cup of water is just as hot, *i.e.* has the same temperature, as the larger quantity, although of course there is a great difference in the quantities of heat the two bodies of water contain. *Temperature depends upon the average kinetic energy of the individual particle, while quantity of heat depends upon the average kinetic energy of the individual particle multiplied by the number of particles, that is, upon the total kinetic energy of all the particles.*

§ 111. Designation of Temperature.—To intelligibly describe a particular temperature it is necessary to state by how much and in what direction (whether higher or lower) it differs from a temperature chosen as a standard, and with which, to understand the description, one must be familiar. It is thus with all descriptions, for example, to describe the position of a point one must, directly or indirectly, state its distance and direction from a known point. To describe the position of a point requires a known point and a unit of length. To describe a temperature requires a *known temperature* and a *unit of difference of temperature.* The temperature which has, for ordinary purposes, been chosen as a standard is the temperature of ice in the act of melting under the pressure of one atmosphere. As you have had an opportunity of observing, this temperature is always the same, and it may be easily obtained. The unit of difference of temperature, to be understood, must, of course, be the difference between two known temperatures, or some fraction of this difference. Different units are used, but each is a fraction of the difference between the temperature at which ice melts and the temperature of the steam from water boiling under the pressure of one atmosphere. The first of these temperatures is called the *freezing point* and the second the *boiling point* of water.

§ 112. Means of ascertaining Temperature.—

We possess a sense by means of which we may observe temperatures directly, and, doubtless, careful training would enable us by this sense to observe with considerable accuracy within a narrow range. For scientific purposes, however, we require means of ascertaining temperatures very accurately through a wide range. A change of temperature produces many other changes in a body, some of which may be readily and very accurately observed, and by observing one of these changes we may draw conclusions regarding the change of temperature. As you have already seen (§ 107), a change of temperature produces a change of volume; and instruments are constructed so that the changes of volume of a certain body may be accurately observed, and hence its changes of temperature accurately inferred. Such an instrument is called a *thermometer*.

§ 113. **Construction of a Thermometer.**—A thermometer generally consists of a glass tube of uniform capillary bore, terminating at one end in a bulb. The bulb and part of the tube are filled with mercury, and the space in the tube above the mercury is usually, but not necessarily, a vacuum. On the tube, or on a plate of metal behind the tube, is a scale to show the height of the mercurial column, and, hence, the volume of the mercury.

§ 114. **Graduation of Thermometers.**—The bulb of a thermometer is first placed in melting ice, and allowed to stand until the surface of the mercury becomes stationary, and a mark is made upon the stem at that point, and indicates the *freezing point*. Then the instrument is suspended in steam rising from boiling water, so that all but the very top of the column is in the steam. The mercury rises in the stem until its temperature becomes the same as that of the steam, when it again becomes stationary, and another mark is placed upon the stem to indicate the *boiling point*. Then the space between the two points found is divided into a convenient number of equal parts called *degrees*, and the scale is extended above and below these points as far as desirable.

Two methods of division are adopted in this country: by one, the space is divided into 180 equal parts, and the

	F.	C.	Abs. temp.	
Water boils............	212°	100°	373°	⎫
Blood heat............	98°	37°	310°	
Max. den. of water....	39·2°	4°	277°	
Water freezes..........	32°	0°	273°	
Mercury freezes........	−37·8°	−38·8°	234·2°	⎬ In centigrade degrees.
No heat...............	−460°	−273°	0°	⎭

Fig. 109.

result is called the *Fahrenheit* scale, from the name of its author; by the other, the space is divided into 100 equal parts, and the resulting scale is called *centigrade*, which means *one hundred steps*. In the Fahrenheit scale, which is generally employed for ordinary household purposes, the freezing and boiling points are marked respectively 32° and 212°. The 0 of this scale (32° below freezing point), which is about the lowest temperature that can be obtained by a mixture of snow and salt, was incorrectly supposed by the inventor to be the lowest temperature attainable. The centigrade scale, which is generally employed by scientists, has its freezing and boiling points more conveniently marked, respectively 0° and 100°. A temperature below 0° in either scale is indicated by a minus sign before the number. Thus, $-12°$ F. indicates 12 Fahrenheit degrees below Fahrenheit's 0 (or 44 Fahrenheit degrees below freezing point). Under F. and C., Fig. 109, the two scales are placed side by side, so as to exhibit at intervals a comparative view. It will be observed that in the construction of the mercury thermometer it is assumed that equal changes of temperature produce equal changes in the volume of mercury. It has been ascertained, by investigations of a character beyond the scope of this book, that this assumption is only approximately true.

§ 115. Conversion from one Scale to the Other. —Since 100 Centigrade degrees = 180 Fahrenheit degrees, 1 Centigrade degree = $\frac{9}{5}$ of a Fahrenheit degree. Hence, to convert Centigrade degrees into Fahrenheit degrees, we multiply the number by $\frac{9}{5}$; and to convert Fahrenheit degrees into Centigrade degrees we multiply by $\frac{5}{9}$. In finding the temperature on one scale that corresponds to a given temperature on the other scale it must be remembered that the number that expresses the temperature on a Fahrenheit scale does not, as it does on a Centigrade scale, express the number of degrees above freezing-point. For example, 52° on a Fahrenheit scale is not 52° above freezing-point, but $52° - 32° = 20°$ above it.

Hence, if you wish to represent a given temperature on the Fahrenheit scale, determine the number of F. degrees the given temperature is from the freezing-point, and then make allowance for the fact that the freezing-point is marked 32° on the F. scale.

Example 1.—How is 13° C. represented on the F. scale?
$$13 \text{ C. degrees} = 13 \times \tfrac{9}{5} = 23\tfrac{2}{5} \text{ F. degrees.}$$
Therefore, the given temperature is $23\tfrac{2}{5}$ F. degrees above freezing, and hence is represented by
$$23\tfrac{2}{5}° + 32° = 55\tfrac{2}{5}° \text{ on the F. scale.}$$

Example 2.—How is 64° F. represented on the Centigrade scale?

64° F. is 32 F. degrees above freezing.
$$32 \text{ F. degrees} = 32 \times \tfrac{5}{9} = 17\tfrac{7}{9} \text{ C. degrees.}$$

Therefore, the given temperature is $17\tfrac{7}{9}$ C. degrees above freezing, and hence is represented by
$$17\tfrac{7}{9}° \text{ on the C. scale.}$$

PROBLEMS

1. The difference between two temperatures is 80 Centigrade degrees. What is the difference in Fahrenheit degrees?

2. When the temperature of a room falls 30 Fahrenheit degrees, how many Centigrade degrees is its temperature lowered?

3. Suppose the temperature of the above room, before the fall, was 68° F., (*a*) what was its temperature after the fall? (*b*) What were the temperatures of the room before and after the fall, according to a Centigrade thermometer?

4. Express the following temperatures of the Centigrade scale in the Fahrenheit scale: 100°; 40°; 56°; 60°; 0°; −20°; −40°; 80°; 150°.

5. Express the following temperatures of the Fahrenheit scale in the Centigrade scale: 212°; 32°; 90°; 77°; 20°; 10°; −10°; −20°; −40°; 40°; 59°; 329°.

§ 116. Air-thermometer.—Prepare apparatus as shown in Fig. 110. A is a glass flask of about one-fourth litre capacity, tightly stopped. Through the stopper extends a glass tube about 60^{cm} long, which also passes through the stopper of a bottle B, partly filled with coloured water. The latter stopper is pierced by a hole *a*, to allow air to pass in and out freely. A strip of paper, C, containing a scale of equal parts, is attached to the tube by means of slits cut in the paper.

Grasp the flask with the palms of both hands, and thereby heat the air in the flask and cause it to expand and in part escape through the liquid in bubbles. When several bubbles have escaped remove the hands, and the air, on cooling, will contract, and the liquid will rise and partly fill the tube.

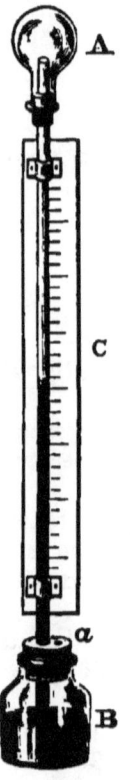

Fig. 110.

By using a large flask and a tube having a small bore, this instrument may be made to indicate clearly very slight changes in temperature. What effect would a change in the atmospheric pressure have on the indications of this thermometer? To what property of air is this due? Observations of other changes which accompany change in temperature have been utilised as a means of inferring temperature.

§ 117. **Testing Extreme Temperatures.**—Mercury boils at 348·8° C. (660° F.), and freezes at −39° C. (−38.2° F.), and therefore cannot be used for indicating temperatures above or below these points. Extremely high temperatures may be tested by the expansion of solids, for example, a rod of platinum, and an instrument used for this purpose is called a *pyrometer*. Alcohol is used in thermometers employed to test extremely low temperatures.

QUESTIONS

1. Wishing to know the temperature of a room, you look at a thermometer which has been for some time in contact with the air of that room. What is it that you actually observe? What do you infer from your observation? There are three substances involved—the mercury, the glass, and the air of the room. Applying the facts learned from experiments already made, from your observation of the *volume* of the *mercury* trace

your inference regarding the *temperature* of the *air* in the room.

2. When testing the temperature of a body by means of a thermometer, what *observation* leads you to infer that the body under examination and the mercury of the thermometer have the *same* temperature?

DIFFUSION OF HEAT

There is always a tendency to *equalisation of temperature*; that is, heat has a tendency to pass from a warmer body to a colder, or from a warmer to a colder part of the same body, until there is an equilibrium of temperature.

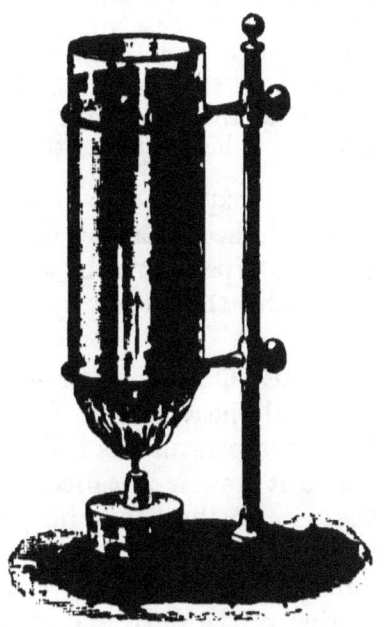

Fig. 111.

If you put your hand in contact with a body at a lower temperature than your hand molecular kinetic energy passes from your hand to the body, and you experience a *sensation* which leads you to say that the body feels cold. If the body is at a higher temperature, your hand gains molecular kinetic energy, and you experience a sensation which leads you to say that the body feels warm. *The intensity of the sensation depends upon the rate at which your hand loses or gains molecular kinetic energy. In other words, it depends upon the rate at which the temperature of your hand falls or rises. The rate at which the temperature of your hand changes will depend, of course, upon the* DIFFERENCE *between the temper-*

ature *of your hand and the temperature of that part of the observed body in immediate contact with it.*

Experiment 1.—Place one end of a wire about 15^{cm} long, in a lamp-flame, and hold the other end in the hand. Apply your fingers to points nearer and nearer to the flame. What do you learn from this experiment?

Experiment 2.—Fill a glass vessel like that in Fig. 111 with water, and stir in a small quantity of fine saw-dust. Apply heat to the bottom and carefully watch the result. By means of a thermometer, from time to time test the temperature of the water at different depths. Do you find much difference between the temperature near the bottom and the temperature near the top? Can you explain the movement which you see taking place in the water? Does this movement aid in the diffusion of the heat throughout the water? Assuming the dynamical theory of heat to be true, how do you explain the diffusion of heat through the wire used in Experiment 1?

These experiments illustrate two quite different ways in which heat may be diffused throughout a body. In the first experiment the heat is said to be diffused by *conduction*, in the second by *convection*.

§ 118. Conduction.—The flow of heat through an unequally-heated body, from places of higher to places of lower temperature, is called *conduction*; the body through which it travels is called a *conductor*. The molecules of the wire in the flame, in Experiment 1 above, have their motion quickened; they strike their neighbours, and quicken their motion; the latter in turn quicken the motion of the next, and so on, until some of the motion is finally communicated to the hand, and creates in it the sensation of heat.

Experiment 1.—Hold wires of different metals of the same length, also a glass tube, a pipe-stem, etc., in the flame, and notice the difference in the time that elapses before the sensation of heat is felt in the different bodies at points a short distance from the flame.

Experiment 2.—Go into a cold room, and place the bulb of a thermometer in contact with various substances in the room; you will probably find that they have the same, or very nearly the same, temperature. Place your hand on the same substances; they appear to have very different temperatures. Try a piece of iron and a block of wood. The iron feels colder than the wood. Keep one hand on the iron and the other on the wood for one minute. Now apply the thermometer to the surfaces lately in contact with your hands. Are they still at the same temperature? You find that the surface of the wood has been raised by contact with your hand to a higher temperature than that of the iron has in the same time. Does this fact explain the difference between the sensations experienced in the two hands? How can you account for the fact itself? Is it possible that the wood is a bad conductor, and hence the part in contact with the hand is quickly warmed, because the heat it receives from the hand is not conducted away to other parts of the block? What experiment can you suggest to determine whether wood conducts heat rapidly or not?

Experiment 3.—Twist together at one end similar wires or strips of iron, copper, brass, etc., 10 or 15^{cm} long, and introduce the twisted ends into a small flame. After a few minutes you can tell approximately the order of their conducting powers, by moving a match along each wire, and seeing how far from the flame it will light. Which do you find to be the best conductor?

You learn that some substances conduct heat much more rapidly than others. The former are called *good conductors*, the latter *poor conductors*. Metals are the best conductors, though they differ widely among themselves.

Experiment 4.—Fill a test-tube nearly full of water, and hold it somewhat inclined (Fig. 112), so that a flame may heat the part of the tube near the surface of the water. Take care that the flame does not strike the tube above the water, or the tube will break. Do you find that the heat is rapidly or slowly transferred to the lower part of the tube? Why are convection currents not set up in this case?

Liquids, as a class, are worse conductors than solids. Gases are much worse conductors than liquids. It is difficult to discover that pure, dry air possesses any conducting power. The poor conducting power of our clothing is due to the poor conducting power of the fibres of the cloth in part, but chiefly to that of the air which is confined by it. (Why is loose clothing warmer than that closely fitting?)

Fig. 112.

Bodies are surrounded with bad conductors, to *retain* heat when their temperature is above that of surrounding objects, and to *exclude* it when their temperature is below that of surrounding objects.

§ 119. **Convection.**—When a hot brick, or a bottle of hot water, is placed at one's feet, heat is also conveyed to the feet. When heat is transferred from one place to another by the *bodily moving* of heated substances, the operation is called *convection*; but this term is rarely applied to solids. Heat does not set up *bodily motion* in solids, but it frequently does in fluids, as you have already seen in the case of water.

Experiment 1.—Fill a small (6 ounce) thin glass flask with boiling hot water, colour it with a teaspoonful of ink, stopper the flask, and, without inverting it, lower it deep into a tub pail, or other large vessel filled with cold clear water. Withdraw the stopper. What takes place? Why does this result follow? What has become of the heat at first stored up in the hot coloured water?

Experiment 2.—Again fill the flask with hot coloured water, stopper, invert, and introduce the mouth of the flask just beneath the surface of a pail of clear cold water. Withdraw the stopper with as little agitation of the water as possible. What happens? Explain.

Experiment 3.—Provide a tightly-covered tin vessel (Fig. 113) and two lamp-chimneys A and B. Near one side of the top of the cover cut a hole a little smaller than the large aperture of chimney B. Near the opposite side of the cover cut a series of holes of about 7^{mm} diameter, arranged in a circle, the circle being large enough to admit a candle without covering the holes. Light the candle, and cover it with chimney A, which should be outside the circle of holes. Fasten both chimneys to the cover with wax. Hold smoking touchpaper C (paper soaked in a solution of saltpetre and then dried) near the top of chimney B. The smoke will enable you to see the directions of the air currents. Describe and explain these currents. Cover the orifice B with the hand. What happens after a short time? Explain.

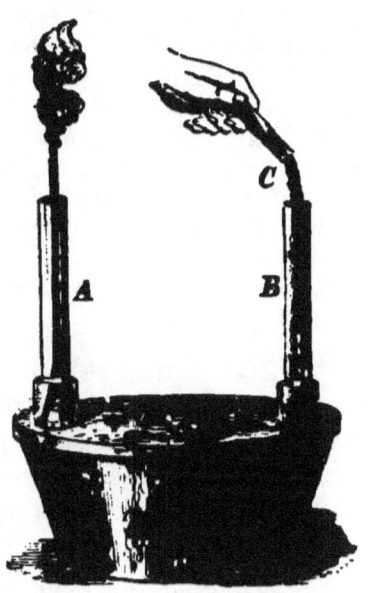

Fig. 113.

The last experiment furnishes an explanation of many familiar phenomena. It explains the cause of chimney-draughts, and shows the necessity of providing a means of ingress as well as egress of air to and from a confined fire. It explains the method by which air is put in motion in winds. It illustrates a method often adopted to ventilate mines. Let the interior of the tin vessel represent a mine deep in the earth, and the chimneys two shafts sunk to opposite extremities of the mine. A fire kept burning at the bottom of one shaft will cause a current of air to sweep down the other shaft, and through the mine, and thus keep up a circulation of pure air through the mine.

Liquids and gases are heated by convection. (Why not solids?) The heat must be applied at the bottom of the body of liquid or gas. (Why not at the top?) There is a still more important method by which heat is diffused, called *radiation*, the method, for example, by which heat reaches us from the sun, which will be treated of under the head of *radiant energy*.

QUESTIONS

1. Why, on a frosty morning, will one's hand freeze to a metallic door-knob rather than to one of porcelain?
2. Why do double windows prevent the escape of heat so much better than single windows?
3. Why, in freezing ice-cream, do we put the freezing mixture in a wooden vessel, and the cream in a metal one?
4. How are safes made fire-proof?
5. Why can one heat water more quickly in a tin than in a china cup?
6. (*a*) How is equalisation of temperature effected in solids? (*b*) In liquids and gases?

§ 120. **Change of Volume accompanying Change of Temperature.**—We have already observed that in solids, liquids, and gases a change of temperature is accompanied by a change of volume. Let us now make some experiments for the purpose of gaining more precise information on this point.

Experiment 1.—Fasten together by rivets, at intervals of 2^{cm}, a strip of sheet iron and a strip of sheet copper, each 15^{cm} long and 2^{cm} wide. Heat this compound strip in the flame of a Bunsen burner or that of a spirit-lamp. What is the result? Do you infer that the iron and the copper expand at the same rate on being heated? Which expands at the greater rate? Cool the compound strip by putting it in ice-water or in a freezing-mixture. What happens? Is this result in accordance with the result when the strip was heated? Make similar experiments with other metals, and learn as much as you can regarding their rates of expansion on being heated.

Experiment 2.—Obtain a glass tube about 15^{cm} long, and having a uniform bore of about 5^{mm}, and close one end by heating in the flame of a Bunsen burner or in that of a spirit-lamp. Partly fill this tube with water, and tie to it a chemical thermometer so that the scale on the thermometer tube may be used to measure the column of water. Place the apparatus in a beaker of ice-water at such a depth that the water column in the tube may be wholly below the surface of the ice-water. Carefully observe and write down the length of the water column in the tube when it has reached the same temperature as the ice-water. Remove the tube and plunge it into a beaker of hot water. When the water column has ceased to rise, and hence has reached the same temperature as the hot water outside, carefully observe and write down its length, as observed by means of the attached thermometer scale; also carefully observe and write down the temperature. What is the length of the water column when in the ice-water, and hence at the temperature of 0° C.? What is its length at the higher temperature? By how many degrees do the two temperatures differ? What is the average increase in length for each degree? Repeat the experiment, using hot water at different temperatures. Do you obtain the same average expansion for each degree in all the experiments? Make similar experiments, using alcohol in the tube instead of water. Do you find that alcohol expands at the same rate as water? By what fraction of its volume at 0° does alcohol expand for each Centigrade degree that its temperature is raised? Do you obtain the same average in all your experiments with alcohol?

The fraction of its volume at 0° C., by which a substance expands when its temperature is raised one Centigrade degree, is called the *coefficient of expansion* of that substance.

Experiment 3.—Fit a perforated cork to a test-tube and insert a glass tube of small bore about 30^{cm} long just reaching through the cork. The cork and tube must fit water-tight. Fill the test-tube and part of the long tube with cold water, excluding all air-bubbles. Place this apparatus in a beaker of ice-water. When the water in the test-tube has ceased to change in volume, and hence has reached the same temperature as the water in the beaker, or 0° C., mark the point at which it stands in the long tube, and remove the ice from the water in the beaker, at the same time introducing a thermometer. As the ice is removed, the water in the beaker will slowly become warmer. Do you find that the water in the test-tube expands as its temperature rises above 0° C.? What does happen? At what temperature do you find the water in the test-tube to have its least volume, and hence its greatest density?

Experiment 4.—Into a graduated glass tube pour about 20^{cm} of kerosene; cool this to 0° C. by putting the tube in ice-water. Drop in some small pieces of dry ice. Observe the total volume, and hence determine the volume of ice introduced. Allow the ice to melt, and observe the volume of water resulting. What is the ratio of the volume of the water to the volume of the ice from which it comes? What is the density of ice?

Experiment 5.—Procure a glass tube about 20^{cm} long, and having a uniform bore of about 1^{mm}. Introduce into it an index of mercury about 5^{mm} long by plunging it into a bottle of mercury, placing the finger firmly over the outer end of the tube, and removing it from the bottle. Move the index to the middle of the tube and close one end by holding it, in a horizontal position, in a gas or alcohol flame. Tie to the tube a chemical thermometer so that its scale may be used to measure the length of the confined air column. Place this apparatus in a horizontal position in a shallow pan filled with ice-water. Note precisely the length of the air column when the thermometer indicates 0° C. Remove the ice and introduce warm water so as to obtain a temperature of 10° C.; note the length of the air column at this temperature. Observe and note the length of the air column at 20° C., at 30° C., and at other temperatures. Before making an observation, tap the tube to facilitate the movement of the mercury index. Fill out the following table:—

Temperature.	Length of Air Column.	Increase in length for 10 degrees.	Average increase in length for each degree Centigrade.
0° C.			
10° ,,			
20° ,,			
30° ,,			
40° ,,			
50° ,,			
60° ,,			
70° ,,			
80° ,,			
90° ,,			
100° ,,			

What do you find to be the *coefficient of expansion* of air ? In these experiments the air has been subject to *constant pressure*. Combining the knowledge gained from these experiments with the fact stated in Boyle's law (§ 98), state the relation that must exist between the temperature and the pressure of a body of gas whose volume is kept constant.

It has been ascertained that the coefficient of expansion is almost exactly the same in the case of all gases.

§ 121. Absolute Temperature.—By Experiment 5 (§ 120) it is found that if a body of air at 0° C. is heated while the pressure remains constant, its volume is increased $\frac{1}{273}$ of the original volume for every degree C. its temperature is raised. But it has been learned (§ 98) that at constant temperature the pressure of a given mass of gas varies inversely as its volume, hence if a body of air is heated while its volume is kept constant, its pressure is increased $\frac{1}{273}$ of the original pressure for every degree its temperature is raised, and therefore at 273° C. its pressure is doubled. If a body of air is cooled below 0° C., while its volume remains constant, its pressure is diminished by $\frac{1}{273}$ of its pressure at 0° for every degree its temperature is lowered ; and, therefore, *if its pressure were to continue to diminish at the same rate* (and in the case of a *nearly perfect gas*, like air, it has been found to diminish almost exactly at the same rate, so far as experiments have been made) at $-273°$ C. its pressure would become nothing. When a gas exerts no pressure on the containing vessel, its molecules must have too little motion to overcome the cohesion between them. Now, the intromolecular cohesion in a gas like air, which is liquefied only at a very low temperature and under great pressure, is believed to be almost *nil*, and hence we may suppose that when it ceases to exhibit any tendency to expand, that is, when it ceases to exert pressure on the containing vessel, its molecules have almost no motion, and the gas is therefore at about its *lowest possible temperature.* For this reason −273° C. is called *absolute zero,* and temperature reckoned from this point is called *absolute temperature.* On this scale all temperatures are, of course, positive. Any particular temperature is designated on the absolute scale by stating the number of degrees this temperature is above absolute zero.

§ 122. Law of Charles.—The relation existing between the volume and the pressure of a given mass of gas may be briefly stated as follows : *At constant pressure the volume of a given mass of gas varies directly as the absolute temperature.* This is called the *Law of Charles.* From this law and from Boyle's law (§ 98) it follows that *at constant volume the pressure of a given mass of gas varies directly as the absolute temperature.*

§ 123. Pressure of a Gas due to the Kinetic Energy of its Particles.—The laws of Boyle and of Charles are both satisfactorily explained by supposing that the pressure of a gas (or its tension or expansive power, as it is sometimes called) is due entirely to the striking of its particles against the surfaces on which the gas is said to press, the impulses following one another in such rapid succession that the effect produced cannot be distinguished from constant pressure. Upon the average momentum of these blows, and upon the number of blows per second on a unit area must, in this case, depend the intensity of the pressure exerted by the gas. According to the dynamical theory of heat, the absolute temperature varies directly as the average kinetic energy of the particles. Imagine the average speed of the particles of a body of gas to be doubled. What effect has this upon the average momentum of the particles ? What effect has it upon the average kinetic energy of the particles ? What effect has it upon the absolute temperature of the gas ? If the volume of the body of gas is kept constant, what effect has doubling the average speed of its particles on the number of blows of these particles per second upon a unit area with which the gas is in contact? What should be the effect, then, upon the pressure exerted by the gas? Imagine the body of gas to be kept at constant temperature and its volume to be doubled. What effect has this upon the average effect of each blow? What effect on the number of blows per second on a unit area? What effect, then, should it have on the pressure?

The foregoing supposition regarding the pressure of a gas is called *the kinetic theory of gases*.

§ 124. Diffusion of Gases.—The kinetic theory of gases explains why gases penetrate into any spaces open to them, and likewise the phenomenon known as the *diffusion of gases* (see § 41). The presence of a gas in a given space only delays the spread of another gas in the same space by collision between the particles of the inter-diffusing gases.

PROBLEMS

1. Find in both Centigrade and Fahrenheit degrees the absolute temperatures at which mercury boils and freezes.

2. At 0° C. the volume of a certain body of gas is 500^{ccm} under a constant pressure; (*a*) what will be its volume if its temperature is raised to 75° C.? (*b*) What will be its volume if its temperature becomes – 20° C.?

3. If the volume of a body of gas at 20° C. is 200^{ccm}, what will be its volume at 30° C.?

4. To what volume will a litre of gas contract if cooled from 30° C. to – 15° C.?

5. One litre of gas under a pressure of one atmosphere will have what volume if, at a constant temperature, the pressure is reduced to 900^g per square centimetre?

6. The volume of a certain body of air at a temperature of 17° C., and under a pressure of 800^g per square centimetre, is 500^{ccm}; what will be its volume at a temperature of 27° C. under a pressure of 1200^g per square centimetre?

7. If the volume of a body of gas under a pressure of 1^k per square centimetre, and at a temperature of 0° C., is 1 litre, at

what temperature will its volume be reduced to 1^{ccm} under a pressure of 200^k per square centimetre?

8. If a cubic foot of coal-gas at 32° F., when the barometer is at 30 in., weighs $\frac{1}{75}$ lb., how much will an equal volume weigh at 68° F. when the barometer is at 29 in.?

9. What is the temperature at the bottom of a pond when ice begins to form at the surface? Where will the water having the greatest density lie?

10. Why are tires made hot when they are to be fitted on carriage-wheels?

11. When a glass stopper is stuck fast in a bottle, it may be loosened by passing a stretched cord once round the neck of the bottle, and rapidly moving the bottle back and forwards so as to produce friction between the cord and the neck of the bottle. Explain this.

§ 125. Fusion and Boiling.

By Experiment 9 (Introductory Experiments) we learned that heat is expended in changing ice to water without raising its temperature. Let us now endeavour to learn something more regarding *change of state*.

Experiment 1.—Melt separately tallow, lard, and beeswax. When partially melted stir well with a thermometer, and ascertain the melting points of each of these substances.

Experiment 2.—Place a test-tube (Fig. 114), half filled with ether, in a beaker containing water at a temperature of 60° C. Although the temperature of the water is 40° below *its* boiling point, it very quickly raises the temperature of the ether sufficiently to cause it to boil violently. Introduce a chemical thermometer[1] into the test-tube, and ascertain the boiling point of ether. After the ether begins to boil does its temperature change?

Fig. 114.

Experiment 3.—Partially fill a flask with water and place it over a Bunsen burner and heat. Carefully watch the temperature until the water has been boiling a short time. What do you

---

[1] A chemical thermometer has its scale on the glass stem, instead of a metal plate, and is otherwise adapted to experimental use.

observe ? Place more burners under the beaker; the water boils more violently; does the temperature rise ? Expose the bulb of the thermometer to the steam near the surface of the boiling water. What temperature does it indicate ?

Experiment 4.—Place in contact the smooth, dry surfaces of two pieces of ice; press them together for a few seconds; remove the pressure, and they will be found firmly frozen together. The ice at the surfaces of contact melts under the pressure, but when the pressure is removed the liquid instantly freezes and cements the pieces together. It is in this manner that snowballs are formed.

NOTE.—If a thermometer is placed in a mixture of ice and water, and the mixture is subjected to great pressure, some of the ice will melt and the temperature will fall; but when the pressure is removed, a portion of the water freezes and the temperature rises. From this we learn that *the melting (or freezing) point of water is very slightly lowered by pressure.* The depression is about $\frac{1}{135}$ of 1° C. for each atmosphere. On the other hand, it is found that *substances which, unlike ice, expand in melting, have their melting points raised by pressure.*

Experiment 5.—Half fill a thin glass flask with water. Boil the water over a Bunsen burner; the steam will drive the air from the flask. Withdraw

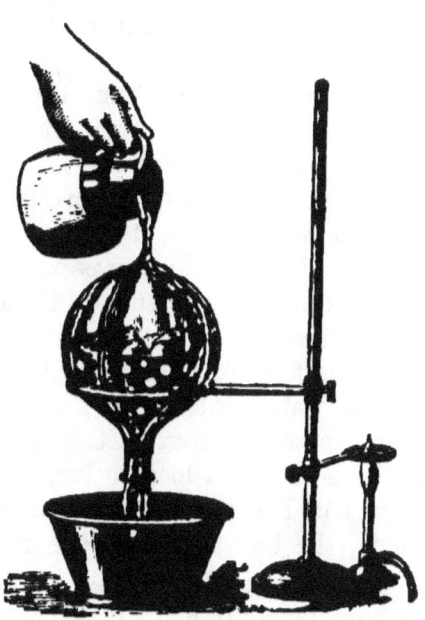

Fig. 115.

the burner, quickly cork the flask very tightly, and plunge the flask into cold water, or invert the flask and pour cold water upon the part containing steam, as in Fig. 115. What is the result ? Pour very hot water on the flask. What is the result ? Now pour on cold water again. What takes place ? Ascertain with your thermometer the temperature of the water in the flask.

Experiment 6.—Put some water in an open beaker and heat it to the boiling point. Stand a thermometer in the water and place beaker and water under the receiver of an air-pump and exhaust the air. What takes place? Does the temperature of the water change? Allow the air to re-enter the receiver, and what follows? What do you observe on the interior surface of the glass receiver of the air-pump? What does this prove? What can be the cause of the water's boiling at a certain temperature in one case and not boiling at the same temperature in the other case? What is the only condition not the same in the two cases, so far as you can see?

Can you now account for the phenomena of Experiment 5? What do these experiments teach?

When high temperature is objectionable, apparatus is contrived for boiling and evaporating in a partial vacuum; as, for instance, in the vacuum-pans used in sugar refineries. As water boils at a temperature lower than 100° C. when the pressure on its surface is less than one atmosphere, so, when the pressure on its surface is more than one atmosphere, it boils at a temperature higher than 100° C.; indeed, the temperature to which water may be raised under the pressure of its own steam is only limited by the strength of the vessel containing it. Vessels made steam tight are often employed to effect a complete penetration of water into solid and hard substances. By this means gelatine is extracted from the interior of bones. In the boiler of a locomotive, where the pressure is sometimes 150 lbs. per square inch above that of the atmosphere, the boiling point rises to about 180° C. (356° F.)

Experiment 7.—Boil some water in which common salt has been dissolved, and carefully note the temperature of the liquid and of the steam just above the surface of the boiling liquid. What do you find? In a similar manner treat the solutions of other solids in water. Make this experiment with a mixture of alcohol and water. Prepare a statement of any general conclusion you reach from your observations. Test the truth of this conclusion in as many ways as you can.

From the foregoing and other similar experiments we

learn the following important laws regarding change of state :—

LAWS OF FUSION AND BOILING

1. *The temperature at which solids melt differs for different substances, but is invariable for the same substance, if the pressure is constant. Substances solidify usually at the same temperatures as those at which they melt.*

2. *After a solid begins to melt, the temperature remains constant until the whole is melted.*

3. *Pressure lowers the melting (or solidifying) point of substances that expand on solidifying, and raises the melting point of those that contract.*

4. *The freezing point of water is lowered by the presence of salts in solution.*

1. *The temperature at which liquids boil differs for different substances, but the temperature of the vapour is invariable for the same substance if the pressure is constant.*

2. *After a liquid begins to boil, the temperature remains constant until the whole is vaporised, provided, of course, that the pressure on its surface is constant.*

3. *Pressure raises the boiling point of all substances.*

4. *The boiling point of water is raised by the presence of salts in solution.*

REFERENCE TABLES

Melting Points

Alcohol	$-130°$ C.	Zinc	about	$425°$ C.
Mercury	$-38·8°$,,	Silver	,,	$1000°$,,
Sulphuric acid	$-34·4°$,,	Gold	,,	$1200°$,,
Ice	$0°$,,	Cast-iron	,,	1050-$1250°$,,
Phosphorus	$44°$,,	Wrought-iron	,,	1500-$1600°$,,
Sulphur	$115°$,,	Iridium (the most infusible metal)		
Tin	about $233°$,,			
Lead	,, $334°$,,	about		$1950°$,,

Boiling Points under a Pressure of one Atmosphere.

Carbon dioxide	$-78°$ C.	Carbon bisulphide		$48°$ C.
Ammonia	$-40°$,,	Alcohol		$78°$,,
Sulphur dioxide	$-10°$,,	Water		$100°$,,
Ether	$35°$,,	Mercury		$350°$,,

Boiling Points of Water at Different Pressures

	Barometer.		Atmospheres.
184° F. . .	16·68 inches	212° F. . . .	1
190° ,, .	18·99 ,,	249·5° ,, . .	2
200° ,, .	23·45 ,,	273·3° ,, . .	3
210° ,, .	28·74 ,,	306° ,, . .	5
212° ,, .	29·92 ,,	356·6° ,, . .	10

The temperature of boiling water varies with the altitude of places, in consequence of the different atmospheric pressure. A difference of altitude of 533 ft., at points not very far from the sea-level, causes a variation of 1° F. in the boiling point.

Boiling Points of Water at Different Altitudes

	Above the Sea-level.	Mean Height of Barometer.	Temperature.
Quito . . .	+ 9,500 ft.	21·53 in.	195·8° F.
Mont Blanc .	15,650 ,,	16·90 ,,	186° ,,
Mount Washington	6,290 ,,	22·90 ,,	200° ,,
London . . .	0 ,,	30 ,,	212° ,,
Dead Sea (below)	− 1,316 ,,	31·50 ,,	214° ,,

§ 126. Distillation.—Apparatus like that represented in Fig. 116 may be easily constructed. The

Fig. 116.

following experiment will be found interesting and instructive.

Experiment.—Half fill the flask A with water coloured with a few drops of ink. Boil the water, and the steam arising will escape through the glass delivery tube BB. This tube is surrounded in part by a larger tube C, called a *condenser*, which is kept filled with cold water flowing from a vessel D through a siphon S, the water finally escaping through the tube E. The steam is condensed in its passage through the delivery tube, and the resulting liquid is caught in the vessel F. What is the character of the liquid caught in the vessel F? What has been effected in this experiment? Upon what does the accomplishment of this result depend? With the same apparatus attempt to separate water from common salt that has been dissolved in it. Do you succeed?

The foregoing apparatus is called a *still*, and the operation is called *distillation*.

If a volatile liquid, such as alcohol, is to be separated from water, can the above apparatus be used? If so, will the alcohol or the water first leave the flask A? A grocer who had watered his whisky too freely attempted to correct his mistake by boiling the mixture in an open pot. What was the result?

§ 127. Evaporation and Condensation.

Experiment 1.—Wet a pane of glass with cold water, and after setting it aside for a few minutes examine it. Is it still wet? What has become of the water? Repeat the experiment, using hot water. Does it take the same time to dry in this case as when cold water was used? Make the experiment again, using ether instead of water. Observe the time it takes the glass to dry in this case. Do you find any difference in this respect between ether and water? What have you previously learned (Experiment 2, § 125) regarding the boiling of ether?

Experiment 2.—Dampen a piece of cotton with water, and hang it outside when the temperature of the air is below the freezing point. The cloth will, of course, freeze stiff. Allow it to remain in this cold atmosphere for several hours. Does the cloth eventually become dry?

Experiment 3.—Hold a dry pane of glass for a moment over an open vessel in which water is boiling. Examine its surface. Is it still dry? How do you reconcile this result with that of Experiment 1? In one case a wet glass becomes dry, in the other a dry glass becomes wet. What difference is there between the conditions surrounding the glass in the one case and those surrounding it in the other? Remove the glass in Experiment 3 away from the neighbourhood of the boiling water. Does it become dry? What surrounding condition have you changed in moving the glass?

Experiment 4.—Place a sheet of wet glass under the receiver of an air-pump and exhaust the air. Does the glass dry more or less quickly than when exposed to the full pressure of the atmosphere?

Vaporisation that takes place quietly and slowly at the surface of a liquid is called *evaporation*. The converse process, or the changing of vapour to liquid, is called *condensation*.

From the foregoing experiments it is learned that the rapidity with which a wet surface becomes dry depends upon the temperature, upon the liquid with which the surface is wet, upon the amount of the vapour of that liquid in the surrounding atmosphere, and upon the pressure upon the wet surface. Doubtless evaporation and condensation are both going on at the same time. If the former exceeds the latter, the surface will become dry. If the two are equal, the surface will remain, so far as we can see, the same. If the latter exceeds the former, the surface will become more wet. Assuming the dynamical theory of heat to be correct, what effect would you expect each of the conditions—temperature, amount of vapour in the surrounding atmosphere, and pressure—to have upon the rate of evaporation? What effect would you expect each to have upon the rate of condensation? Are the results of your experiments in accordance with these expectations? You must bear in mind that in each case you have observed not the result of the evaporation alone, or of the condensation alone, but the difference of these results.

The excess of condensation over evaporation may show itself not only in the appearance of liquid on surfaces in contact with the vapour, but also in the appearance of minute drops of liquid in the midst of the vapour, as in the case of *fog* or *cloud*.

§ 128. Dew Point.—When a space contains such an amount of water-vapour, whether it contains other gases or not, that its temperature cannot be lowered without some of the water being precipitated in the form of a liquid, the space is said to be *saturated*, and the temperature is called the *dew point*. The form in which the condensed vapour appears is, according to its location, *dew*, *fog*, or *cloud*. The atmosphere is said to be *dry* or *humid*, according as the difference between the dew point and the temperature of the atmosphere is great or little.

QUESTIONS

1. Why does our breath usually produce a cloud in winter and not in summer?
2. (a) If air at 0° is warmed to 20° C., how will its dryness be affected? (b) What effect would such warmed air have on wet clothes?
3. If saturated air at 20° is blown into a cellar where the temperature is 10°, what will happen?
4. What is the cause of the general complaint of dryness of air in rooms heated by stoves or furnaces?
5. Does a given mass of air in such a room contain less water-vapour than an equal mass of cold outdoor air at the same time?
6. Find the dew point of your school laboratory by the following experiment: Take a bright nickel-plated or silver cup; pour into it a small quantity of tepid water. Place in the water the bulb of a chemical thermometer. Gradually reduce the temperature of the water by stirring into it ice-water until you discover a slight dimness of the lustre of that portion of the outside of the cup next the water. If the ice-water does not reduce the temperature sufficiently, add ice, keeping the mixture briskly stirring. If the ice does not answer, pour out some of the water and sprinkle salt on the ice, keeping the bulb of the thermometer in the remaining water. Note the temperature of the water at the instant that the first mist or dimness appears on the cup. Wait until the dimness or mist disappears, and note the temperature of the water when the last disappears. Take the mean of the two temperatures for the dew point.

7. In this experiment why is a cup having a *bright, smooth surface* used? What reason have you to infer that the temperature of the water is nearly the same as that of the atmosphere in contact with the outside of the cup? Why are two observations made, and the mean of the two taken as the dew point?

8. Why is a fog usually found about an iceberg?

9. What examples of evaporation or of condensation have you observed in nature?

§ 129. Measurement of Heat.—Hitherto we have experimented for the purpose of learning facts regarding *temperature*, let us now turn our attention for a time to the question of *quantity of heat*. For this purpose, as in the case of all measurements, we shall require a unit. The heat unit generally adopted is *the quantity of heat required to raise the temperature of one kilogramme of water from 0° to 1° C.* This unit is usually called a *calorie*.

Let us find the quantity of heat required to melt a kilogramme of ice without changing its temperature.

Experiment 1.—Mix 1^k of water at 0° with 1^k at 20°; the temperature of the mixture becomes 10°. The heat that leaves 1^k of water when it falls from 20° to 10° is just capable of raising 1^k of water from 0° to 10°.

From this experiment you learn this important truth. *A body, in cooling, gives to surrounding bodies as much heat as is required to restore its own temperature.*

Experiment 2.—Put a kilogramme of snow or pounded ice at 0° C. into a kilogramme of water at 100° C., and rapidly stir the mixture until the snow is melted. Observe the temperature at the moment the snow is melted. In making this experiment wrap the vessel in flannel to prevent communication of heat to or from the atmosphere.

Let $a°$ be the temperature thus observed. One kilogramme of water has been cooled $(100 - a°)$; one kilogramme of ice has been melted, and the resulting kilogramme of water has been warmed $a°$. Therefore $(100 - 2a)$ calories have been required to melt one kilogramme of ice.

Repeat the experiment with different weights of water and of snow. Taking the average of the conclusions deduced from your experiments, how many calories do you find are required to melt one kilogramme of ice?

Next, let it be required to find the amount of heat that disappears during the conversion of 1^k of water into steam.

Experiment 3.—Place 1^k of water at 0° C. in a beaker, and heat the same with a Bunsen burner. Note the time that it takes to raise the temperature of the water from 0° C. to 100° C. Allow the water to boil for ten or twenty minutes, and, by weighing or measuring the water remaining, find how much has been converted into steam. In making this experiment use a tall beaker, so that no *liquid water* may be thrown out by the boiling.

How many calories must the kilogramme of water receive to raise its temperature from 0° to 100°? How many calories does it receive per minute? Assuming that it receives heat from the flame at a uniform rate, how many calories does it receive *after it begins to boil*? How much of the water is changed to steam by this heat? At this rate, how many calories of heat will change a kilogramme of boiling hot water to steam?

Repeat this experiment, and find the average of the results.

Experiments made by more accurate but more complicated methods than the above give the following results:—

1. *The amount of heat that disappears, or is expended, in the melting of one kilogramme of ice is 80 calories.*

2. *The amount of heat that disappears, or is expended, in the conversion of one kilogramme of water into steam is 537 calories.*

If your experiments are carefully made, and your calculations based thereon are correct, you should reach conclusions differing very little from these.

§ 130. Latent Heat.—Inasmuch as none of the heat applied during the melting of ice and the conversion of water into steam raises the temperature of the body to which it is applied, the question arises, *What does the heat do?* Again, *Why is not ice instantly converted into water on reaching the melting-point, and water instantly converted into steam on reaching the boiling-point?*

According to the dynamical theory of heat, the answer to the first question is, All of the heat applied in melting ice is expended in doing *interior work*, as it is called. The molecules that were firmly held in their places by molecular forces are now moved from their places, and so work is done against these forces, just as work is done against gravity when a weight is lifted. In the conversion of water into steam a similar action goes on; the heat is expended in separating the molecules still further, all except the small fraction used in overcoming atmospheric pressure. Heat, the energy of motion, in both instances does important work, and is thereby transformed into the energy of position, or potential energy,—energy of the same kind as that of a raised weight.

The answer to the second question is, The amount of work done in both instances is great, as shown by the amount of heat expended in doing the work; 80 calories per kilogramme of ice being required in the first instance, and 537 calories per kilogramme of water in the second; hence it requires a long time to acquire the requisite amount of heat. It is fortunate that it takes a large quantity of heat to melt ice; otherwise, on a single warm day in winter, all the ice and snow would melt, creating most destructive freshets. The heat which disappears in melting and boiling is generally, but with our present knowledge of the subject, rather objectionably, called *latent* (hidden) *heat*. The error consists in calling that heat which has ceased to be heat, *i.e.* has ceased to be *molecular motion*. If we should agree to use the word heat to signify molecular energy of position as well as of motion, the expression "latent heat" would not be inappropriate, since molecular potential energy is not discernible by means of our heat sense.

§ 131. Artificial Means of Lowering Temperature.—The fact that heat must be expended because work is done, in the conversion of solids into liquids and liquids into vapours, and in the simple expansion of gases (when external work is done in this expansion), is turned to practical use in many ways for the purpose of obtaining a low temperature. The following experiments will illustrate :—

Experiment 1.—Prepare a mixture of 2 parts by weight of pulverised ammonium nitrate and 1 part of ammonium chloride, and dissolve in 3 parts of water (not warmer than 10° C.), stirring the same, while dissolving, with a small test-tube containing a little cold water. What is the result ? What temperature is indicated by a thermometer placed in the mixture ?

One of the most common freezing mixtures consists of 3 parts snow or broken ice and 1 part of common salt. The affinity of salt for water causes a liquefaction of the ice, and the resulting liquid dissolves the salt, both operations requiring heat.

Experiment 2.—Fill the palm of the hand with ether; the ether quickly evaporates, and what is the result ?

Experiment 3.—Place water at about 10° C. in a thin porous cup, such as is used in the Grove's battery, and introduce the bulb of a thermometer ; although the experiment be conducted in a warm room, the large surface exposed by means of the porous vessel will so hasten evaporation that in the course of fifteen minutes there will be a very sensible fall in temperature.

Fig. 117.

Experiment 4.— Cover closely the bulb of an air thermometer (Fig. 117) with thin muslin, and partly fill the stem with water. Let one person slowly drop ether on the bulb while another briskly blows the air charged with vapour away from the bulb with a bellows. (Why ?) Do you observe any evidence of a lowering of temperature? Can you freeze the water in the stem of the air thermometer?

The following plan has been adopted for keeping fresh meat frozen on shipboard even in the tropics. Air is forced under great pressure into strong iron cylinders. These cylinders and the contained air (which, of course, has been greatly heated in being thus compressed) are cooled to the temperature of the sea-water by being kept in contact with it for a time. When thus cooled the compressed air is allowed to escape into the compartment containing the fresh meat, driving out as it enters the air already there. The work done in driving out this air requires the expenditure of so much heat that the expanding air is readily cooled far below the freezing point.

By such a contrivance as this it is as possible to keep a house cool in summer as it is to keep it warm in winter by the burning of coal. Indeed, it is strange that in warm countries those able to bear the expense (which need not be very great) have not made greater use of this means of keeping their houses cool during the hot months.

QUESTIONS

1. Why do we bathe the fevered forehead with alcohol and water?
2. How does perspiration contribute to our comfort?
3. Why do we fan ourselves?
4. Why does a windy day seem colder to us than a still day, although the temperature is the same on both days?
5. How does sprinkling a floor cool the air of a room?

§ 132. Solidification of Liquids and Condensation of Vapours.

Having learned that heat disappears in the processes of fusion and vaporisation, the question naturally arises, Does heat appear in the processes of solidification and condensation?

Experiment 1.—Boil about ½ litre of water in a glass flask, and add, slowly, pulverised sodium sulphate until the boiling water refuses to dissolve more (hot water will dissolve about twice its weight of this substance). Then set the hot solution in a place where it will not be disturbed, and let it stand for about 24 hours, that it may acquire the temperature of the room. Thrust the bulb of a thermometer into the solution, and at the same time drop in a lump of sodium sulphate; solidification instantly sets in, and in a few seconds the liquid mass will be almost wholly replaced by crystals. At the same time, what change takes place in the temperature? What does this change prove?

Experiment 2.—Place water at about 10° C. in a bottle, and introduce a thermometer. Surround the bottle with a snow and salt freezing mixture, in which a thermometer is also placed. From what you have already learned, what transference of heat are you sure is taking place so long as the contents of the bottle and the freezing mixture have different temperatures? What is the temperature of the freezing mixture? At what temperature does the water begin to freeze? Does it become any colder after it begins to freeze? Is it while freezing as cold as the mixture of snow and salt outside? Have you any evidence that the water while freezing is giving up heat to the freezing mixture without its own temperature being lowered in consequence? Is heat produced by the solidification of the water? After the water is all frozen does its temperature change?

Experiment 3.—Arrange apparatus as in Fig. 118.

Fig. 118.

Wrap flannel about the thin glass beaker C and pour into it a known weight of cold water, say 500 grammes. Place a laboratory burner under A, and when a strong jet of steam issues from the delivery tube B let it enter the water in the beaker C, having first carefully noted the temperature of that water.

Carefully and constantly stir the water in the beaker with a thermometer. When the temperature has risen 20° or 30°, carefully note the temperature and remove the delivery tube from the beaker. Weigh the water now in the beaker. From your observations prepare answers to the following questions: What weight of water was put into the beaker at first? By how much has its temperature been changed? How many calories of heat are required to produce that change? What weight of steam has condensed in the beaker? How much heat has the water resulting from this condensation given out in cooling from 100° C. to the temperature of the beaker at the close of the experiment? How much heat must have been produced by the condensation of the steam? At this rate how much heat is produced by the condensation of one kilogramme of steam? How does this compare with the quantity of heat already found necessary (§ 129) to change one kilogramme of water at 100° C. to steam at the same temperature? In this experiment, what purpose do you find is served by introducing the large piece of tubing in the delivery-tube?

From the last and other similar experiments it is proved that *heat that is consumed in liquefying solids, and in vaporising liquids, is always restored when the reverse change takes place.* Farmers well understand that water in freezing gives out a great deal of heat,—at a low temperature, it is true, but still high enough to protect vegetables which freeze only when considerably colder than melting ice. The fact that steam in condensing generates a large amount of heat, is turned to practical use in heating buildings by steam.

The following tables embody the results of experiments made by Regnault, Andrews, and others:—

Substance.	Latent Heat of equal weights of Liquids (Water=1).
Water	1·000
Phosphorus	0·063
Sulphur	0·118
Nitrate of Soda	0·794
Nitrate of Potassa	0·598
Tin	0·179

Substance.	Latent Heat of equal weights of Liquids (Water=1).
Bismuth	0·159
Lead	0·067
Zinc	0·355
Cadmium	0·172
Silver	0·266
Mercury	0·035

Substance.	Latent Heat of equal weights of Vapours (Steam=1).
Water	1·000
Wood spirit	0·492
Alcohol	0·378
Ether	0·169
Bisulphide of carbon	0·162
Oxalic ether	0·136
Formic ether	0·196
Acetic ether	0·173
Iodide of ethyl	0·087
Iodide of methyl	0·086
Bromine	0·085
Perchloride of tin	0·057
Formiate of methyl	0·219
Acetate of methyl	0·206
Terchloride of phosphorus	0·096

It will be seen that more heat is expended in melting a kilogramme of ice than in melting a kilogramme of any other substance. Also more heat is required to vaporise a kilogramme of water than is required to vaporise a kilogramme of any other substance. What effect has the great latent heat of water upon the thickness of the ice on our lakes in winter? What effect has the great latent heat of steam upon the rate of evaporation at the earth's surface? What effect upon the rate of condensation, as in a fall of rain?

§ 133. **Specific Heat.**— Let us ascertain whether the same quantity of heat is required to change, to the same extent, the temperature of equal weights of different substances.

Experiment.—Take (say) 300ᵍ of sheet lead, and make a loose roll of it, and suspend it by a thread in the steam just above the surface of boiling water for about five minutes, that it may acquire the same temperature (100° C.) as the steam. Remove the roll from the steam, and immerse it as quickly as possible in 300ᵍ of water at 0°, contained in a thin glass beaker well wrapped in flannel, and introduce the bulb of a thermometer. Note the temperature of the water when it ceases to rise. Move the lead up and down in the water by means of the suspension cord, so as to bring the lead and the water to the same temperature as quickly as possible. From our observations in previous experiments we may be sure that the lead has lost the same amount of heat that the water has gained. By how many degrees has this quantity of heat changed the temperature of the 300 grammes of water? By how many degrees has its loss lowered the temperature of the 300 grammes of lead? Has the lead the same *capacity for heat* as an equal weight of water? What is the ratio of the capacity for heat of the lead to the capacity for heat of an equal weight of water? Find this ratio in the case of zinc, of copper, of iron, etc.

The ratio of the capacity for heat of a substance A to the capacity for heat of an equal weight of water is called the *specific heat* of the substance A. It has been found that the specific heat of the same substance is not exactly the same at all temperatures, nor is the specific heat of a solid usually the same as that of the same substance in liquid or in a gaseous condition. The following tables embody facts experimentally determined by Dulong and Petit and others:—

Substance.	Mean Specific Heat.	
	Between 0° and 100° C.	Between 0° and 300° C.
Iron	0·1098	0·1218
Mercury	0·0330	0·0350
Zinc	0·0927	0·1015
Antimony	0·0507	0·0549
Silver	0·0557	0·0611
Copper	0·0949	0·1013
Platinum	0·0355	0·0355
Glass	0·1770	0·1990

| | Specific Heat. | | |
Substance.	Solid.	Liquid.	Gaseous.
Water	0·5040	1·0000	0·4805
Bromine	0·0833	0·1060	0·0555
Tin	0·0562	0·0637	...
Iodine	0·0541	0·1082	...
Lead	0·0314	0·0402	...
Alcohol	...	0·5475	0·4534
Bisulphide of carbon	...	0·2352	0·1569
Ether	...	0·5290	0·4797

Water requires more heat to warm it, weight for weight, and gives out more in cooling through a given range of temperature, than any substance except hydrogen. The quantity of heat that raises the temperature of a kilogramme of water from 0° to 100° C. would raise the temperature of a kilogramme of iron from 0° to 800° or 900° C., or above a red heat; conversely, a kilogramme of water, in cooling from 100° to 0° C., gives out as much heat as a kilogramme of iron in cooling from about 900° to 0° C.

§ 134. **Causes of Difference in Capacity for Heat.**—Of the whole quantity of heat imparted to a solid or liquid body only a part goes to increase the heat of the body, and thereby to raise its temperature; the remainder performs *interior work* in overcoming cohesion between the molecules of the body, and in forcing them to take up new positions. The greater the portion of heat consumed in interior work upon a body the less there is left to raise its temperature, and consequently the greater its capacity for heat. Again, considering that portion of heat which does raise the temperature, since the temperature, according to the dynamical theory of heat, depends upon the average kinetic energy of each particle (§§ 108, 109), it is evident that the quantity of heat required to raise the temperature of a unit mass of a substance 1°, is greater the greater the number of particles in the unit mass. Thus, much more heat is required to raise the temperature of a pound of water 1° than to raise that of a pound of lead 1°; (1) because more *interior work* is done in the water than in the lead, and (2) because *there are more particles in a pound of water than in a pound of lead.* The study of chemistry leads to the conclusion that there is between the atomic weight of an element and its capacity for heat such a connection as that indicated by the dynamical theory of heat.

QUESTIONS AND PROBLEMS

1. How much heat is required to change 100^k of ice at 0° into steam at 100° C. ?

2. (*a*) 1000^k of steam at 100° C. is conveyed by pipes through a building, and the water resulting from its condensation returns to the boiler at a temperature of 80°; how much heat is given out in the building? (*b*) The same quantity of heat would raise the temperature of how many kilogrammes of water from 0° to 100° ?

3. 50^k of water at 100° will melt how many pounds of ice at 0° C. ?

4. How much heat is required to change 1^k of ice at $-10°$ to water at 10° C. ?

5. (*a*) Apply the same quantity of heat to equal weights of ice and of water, each at a temperature of 0° C.; when the latter reaches the boiling point what will be the temperature of the former? (*b*) Why will not both have the same temperature?

6. What effect on the temperature of the air has the freezing of the water of lakes and other bodies of water?

7. If 1^k of iron at 100° is immersed in 1^k of water at 0° C., what will be the resulting temperature?

8. What is the specific heat of a substance, 1^k of which at 100°, when put into 1^k of water, at 0° raises its temperature to 5° C. ?

9. 50^k of mercury at 80° will melt what weight of ice at 0° C. ?

10. Why is hot *water* in bottles often used to warm beds in preference to other substances?

11. If there were no water on the earth, why would the

difference in temperature between day and night, and between summer and winter, far exceed what it is now?

12. Why are places in vicinity of water less subject to extremes of heat and cold than places inland?

13. In high latitudes, why have the western coasts of the continents far milder climates than the eastern coasts in the same latitude?

14. Where does the rain that falls in Scotland come from? Where was the vapour from which it comes formed? In connection with this vapour and its condensation, where was heat expended? Where is heat produced?

§ 135. **Mechanical Equivalent of Heat.**—If the dynamical theory of heat is correct, there evidently must

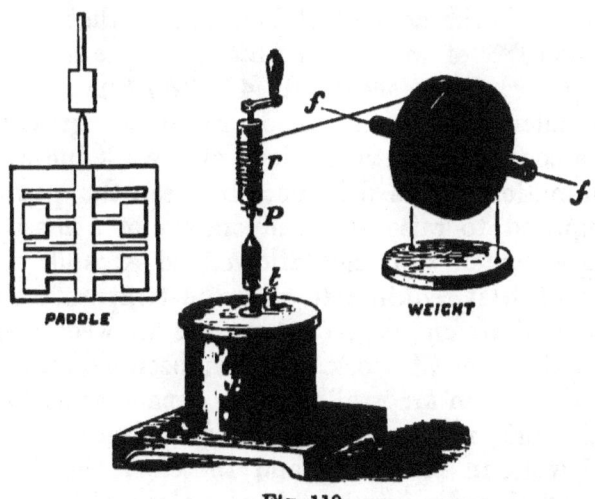

Fig. 119.

be some definite relation between the unit of heat and the unit of mechanical energy. Dr. Joule of Manchester made numerous and laborious experiments, extending over a period of seven years, for the purpose of determining this connection. He caused a paddle-wheel to revolve in water by means of a falling weight attached to a cord wound around the axle of a wheel, as represented in Fig. 119. The resistance offered by the water to the motion of the paddles was the means by which the mechanical motion of the weight was converted into heat, which raised the temperature of the water. Taking a body of a known weight, *e.g.* 80^k, he raised it a measured distance, *e.g.* 53^m high; by so doing 4240^{kgm} of work were performed upon it, and consequently an equivalent amount of energy was stored up in it ready to be converted, first into mechanical motion, then into heat. He took a definite weight of water to be agitated, *e.g.* 2^k, at a temperature of 0° C. After the descent of the weight, the water was found to have a temperature of 5° C.; consequently the 2^k of water must have received 10 units of heat (careful allowance being made for all losses of heat), which is the amount of heat-energy that is equivalent to 4240^{kgm} of work, *or 1 unit of heat is equivalent to* 424^{kgm} *of work* (more accurately $423·985^{kgm}$).

The determination thus made by Joule agrees closely with the connection between heat and work indicated by an examination of the difference between the quantity of heat required to raise the temperature of a mass of air one degree when it is not allowed to expand, and the quantity of heat required to raise the temperature of the same mass of air one degree when it is allowed to expand so as to do external work. It has been experimentally proved that when air is allowed to expand so as to do no external work, no heat is expended. Hence air does no internal work in expanding, and therefore the difference of heat mentioned above is the equivalent of the external work performed.

One of the most important discoveries in science is that of the *equivalence of heat and work*; that is, that *a definite quantity of mechanical work can always produce a definite quantity of heat; and conversely, this heat, if the conversion be complete, can perform the original quantity of work.*

§ 136. Correlation and Conservation of

Energy.—The proof of the facts just stated was one of the most important steps in the establishment of the grand twin conceptions of modern science. (1) That *all kinds of energy are so related to one another that energy of any kind can be changed into energy of any other kind,*—known as the doctrine of CORRELATION OF ENERGY; (2) that *when one form of energy disappears, an exact equivalent of another form always takes its place, so that the sum total of energy in the universe is unchanged,*—known as the doctrine of CONSERVATION OF ENERGY. These two principles constitute the corner-stone of physical science.

§ 137. The Conversion of Mechanical Energy into Heat.—The conversion of mechanical energy into heat is a very simple matter. Indeed, the difficulty is not so much to produce this conversion as to avoid it, and to this intent we use lubricants to diminish the friction of machinery as much as possible.

§ 138. The Energy stored in Coal.—In our coal beds and in the atmosphere nature has stored up a great fund of energy in the form of potential energy of chemical affinity, the particles of the coal being separated from and having a great tendency to unite with the particles of oxygen in the atmosphere. It is a great practical problem with us to convert this store of energy into the energy of mechanical motion for use in our factories, steamships, and railway trains. The only plan yet adopted has been to convert this energy into heat by burning the coal, and then to convert the heat into mechanical motion.

§ 139. Conversion of Heat into Mechanical Energy.—Various contrivances have been employed for the conversion of heat into work, such as the boiler and steam engine, the hot air engine, and the gas engine. Owing to the fact that ordinary temperature at the earth's surface is so far above absolute zero (§ 121), it is impossible, *by any contrivance*, to convert all of a given quantity of heat into mechanical work. For example, if steam is taken into an engine at 150° C. (423° Abs.), and is exhausted at 20° C. (293° Abs.), and the engine is *perfect*, so that none of the heat absorbed by it is dissipated, but is all converted into mechanical work, the heat converted into mechanical work is only $\frac{423-293}{423}$, or $\frac{130}{423}$ of the quantity of heat supplied to the engine. We are, in fact, in the position of a man wishing to utilise the energy of a waterfall, who, from circumstances which he cannot possibly control, cannot set his water-wheel at the foot of the fall, but must set it near the top, and therefore must lose the energy of that part of the fall below his wheel. As will be seen, the *efficiency* of any heat engine is increased by introducing the heat at a higher temperature, or by discharging it at a lower temperature. The ordinary temperature at the earth's surface is such that the temperature of discharge cannot be lowered much beyond the limit already attained. The temperature of introduction is being increased as our artisans advance in the mechanical skill necessary for the production of machinery capable of withstanding high temperature and great pressure.

§ **140. The Steam Engine.**—The modern steam engine consists essentially of an arrangement by which steam from a boiler is conducted to both sides of a piston alternately ; and then, having done its work in driving the piston to and fro, is discharged from both sides alternately, either into the air or into a condenser. The diagram in Fig. 120 will serve to illustrate the general features and the operation of a steam engine. The details of the various mechanical contrivances are purposely omitted, so as to present the engine as nearly as possible in its simplicity.

In the diagram, B represents the *boiler*, F the *furnace*,

S the *steam pipe* through which steam passes from the boiler to a small chamber VC, called the *valve chest.* In this chamber is a *slide valve* V, which, as it is moved to and fro, opens and closes alternately the passages M and N leading from the valve chest to the *cylinder* C, and thus admits the steam alternately to each side of the *piston* P.

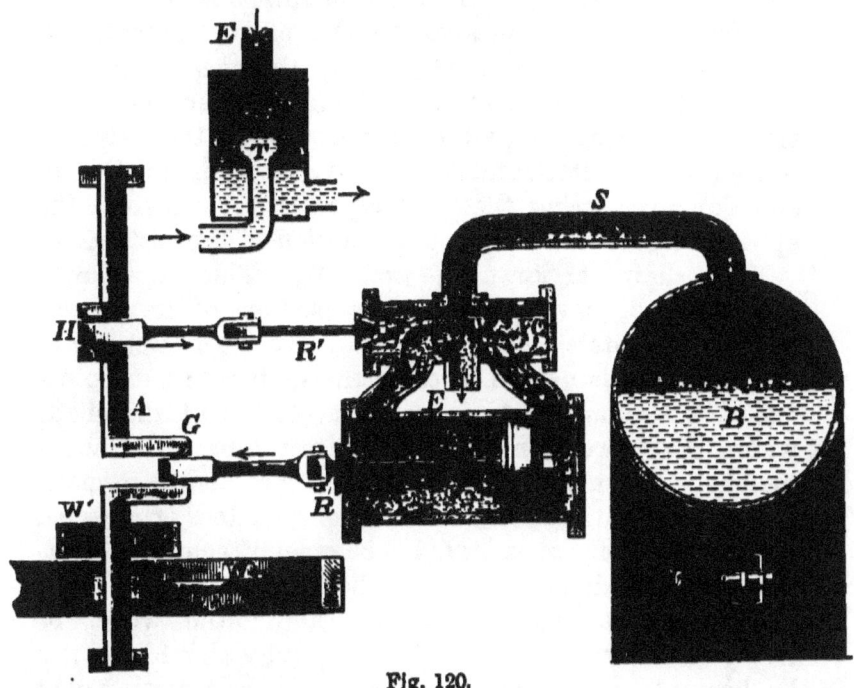

Fig. 120.

When one of these passages is open the other is always closed. Though the passage between the valve chest and the space in the cylinder on one side of the piston is closed, thereby preventing the entrance of steam into this space, the passage leading from the same space is open through the interior of the valve so that steam can escape from this space through the *exhaust pipe* E. Thus, in the position of the valve represented in the diagram, the passage N is open, and steam entering the cylinder at the top drives the piston in the direction indicated by the arrow. At the same time the steam on the other side of the piston escapes through the passage M and the exhaust pipe E. While the piston moves to the left, the valve moves to the right, and eventually closes the passage N leading from the valve chest, and opens the passage M into the same, and thus the order of things is reversed.

Motion is communicated by the piston through the *piston rod* R to the *crank* G, and by this means the *shaft* A is rotated. Connected with the shaft by means of the crank H, is a rod R', which connects with the valve V, so that as the shaft rotates, the valve is made to slide to and fro, and during the greater part of the stroke in the opposite direction to that of the motion of the piston.

The shaft carries a *fly-wheel* W. This is a large, heavy wheel, having the larger portion of its weight located near its circumference; it serves as a reservoir of energy which is needed to carry the shaft past two points (called the *dead points*) in each revolution of the shaft, where the power communicated directly by the steam is ineffectual in moving the shaft. It also assists to make the rotation of the shaft and all other machinery connected with it uniform, so that sudden changes of velocity resulting from sudden changes of the driving power or of resistances are avoided. (Why should the wheel be heavy? Why should it be large? Why should the rim be heavy? See § 58.) By means of a belt passing over the wheel W' motion may be communicated from the shaft to any machinery desirable.

§ 141. Condensing and Non-Condensing Engines.[1]—Sometimes steam, after it has done its work in the cylinder, is conducted through the exhaust pipe to a chamber Q called a *condenser*, where, by means of a spray of cold water introduced through a pipe T, it is suddenly

[1] The terms, *low pressure* and *high pressure* engines, are not distinctive as applied to engines of the present day.

condensed. This water and the condensed steam must be pumped out of the condenser by a special pump called technically the *air-pump*; thus a partial vacuum is maintained. Such an engine is called a *condensing engine*. The advantage of such an engine is obvious, for, if the exhaust pipe, instead of opening into a condenser, communicates with the outside air as in the *non-condensing engine*, the steam is obliged to move the piston constantly against a resistance arising from atmospheric pressure of 15 pounds for every square inch of the surface of the piston. But in the condensing engine no resistance arises from atmospheric pressure, and so with a given steam pressure in the boiler the effective pressure on the piston is considerably increased; hence, condensing engines are usually more economical in their working. The efficiency of an engine is also increased by so arranging the valves that the supply of steam is shut off from the cylinder before the stroke of the piston is finished, thus allowing the steam already introduced to complete the stroke by its own expansion. By this contrivance the steam is cooled within the engine, and the heat it thus loses is converted into mechanical work.

By the best combination of furnace, boiler, and engine yet constructed, mechanical energy has been obtained at the rate of one horse-power per $1\frac{1}{2}$ lbs. of coal consumed per hour. Even in this case, only about $\frac{1}{8}$ of the chemical potential energy of the coal and oxygen is eventually realised as mechanical energy.

§ 142. **Origin of Animal Heat and Muscular Motion.**—The plant finds its food in the air (principally the carbon dioxide in the air) and in the earth, in the condition of a fallen weight; but, by the agency of the sun's radiation, work is performed upon this matter during the growth of the plant; potential energy is stored in the plant,—the weight is drawn up. The animal now finds its food in the plant, appropriates the energy stored in the plant, and converts it into energy of motion in the form of heat and muscular motion. The plant, then, may be regarded as a machine for converting energy of motion received from the sun into potential energy; the animal, as a machine for transforming it again into the energy of motion.

§ 143. **The Sun as a Source of Energy.**—Not only is the sun the source of the energy exhibited in the growth of plants, as well as of the muscular and heat energy of the animal, but it is the source, directly or indirectly, of very nearly all the energy employed by man in doing work. Our coal-beds, the results of the deposit of vegetable matter, are vast storehouses of the sun's energy, rendered potential during the growth of the plants many ages ago. Every drop of water that falls to the earth, and rolls its way to the sea, contributing its mite to the unbounded water-power of the earth, and every wind that blows, derives its power directly from the sun.

If a man were to make use of the ocean tides for driving machinery he would be using energy derived from what source? By the friction of the tides against the coast the water and land are warmed. What is the source of this energy? Is this source inexhaustible? Is it increased itself from any other source? If not, what must be the effect of this continual call upon it? Is any similar effect to be found outside the earth?

QUESTIONS

1. What kind of engine (*i.e.* condensing or non-condensing) is that which produces loud puffs? What is the cause of the puffs?
2. Why does the temperature of steam suddenly fall as it moves the piston?
3. What do you understand by a ten horse-power steam-engine?
4. Upon what does the power of a steam-engine depend?

5. Is the compound engine a condensing or a non-condensing engine? Which is the locomotive engine?

6. The area of a piston is 500 square inches, and the average unbalanced steam pressure is 30 pounds per square inch; what is the total effective pressure? Suppose that the piston travels 30 inches at each stroke, and makes 100 strokes per minute, allowing 40 per cent for wasted energy, what power does the engine furnish, estimated in horse-powers?

THE END

Printed by R. & R. CLARK, *Edinburgh.*

Some of Messrs. Macmillan & Co.'s Books on Physics.

ON SOUND AND ATMOSPHERIC VIBRATIONS. By Sir G. B. AIRY, K.C.B., formerly Astronomer-Royal. With the Mathematical Elements of Music. Crown 8vo. 9s.

MECHANICAL THEORY OF HEAT. By R. CLAUSIUS. Translated by W. R. BROWNE, M.A. Crown 8vo. 10s. 6d.

AN INTRODUCTION TO THE THEORY OF ELECTRICITY. By LINNÆUS CUMMING, M.A., Assistant Master at Rugby. Illustrated. Crown 8vo. 8s. 6d.

A TEXT-BOOK OF THE PRINCIPLES OF PHYSICS. By ALFRED DANIELL, D.Sc. Illustrated. Second Edition, revised and enlarged. 8vo. 21s.

ELECTRIC LIGHT ARITHMETIC. By R. E. DAY, Evening Lecturer in Experimental Physics at King's College, London. Pott 8vo. 2s.

ILLUSTRATIONS OF THE C. G. S. SYSTEM OF UNITS WITH TABLES OF PHYSICAL CONSTANTS. By J. D. EVERETT, F.R.S., Professor of Natural Philosophy, Queen's College, Belfast. New Edition. Extra fcap. 8vo. 5s.

AN ELEMENTARY TREATISE ON SPHERICAL HARMONICS, and Subjects connected with them. By Rev. N. M. FERRERS, D.D., F.R.S., Master of Gonville and Caius College, Cambridge. Crown 8vo. 7s. 6d.

THE THEORY AND PRACTICE OF ABSOLUTE MEASUREMENTS IN ELECTRICITY AND MAGNETISM. By A. GRAY, F.R.S.E., Professor of Physics, University College, Bangor. Two Vols. Crown 8vo. Vol. I. 12s. 6d. [Vol. II. *In the Press.*

ABSOLUTE MEASUREMENTS IN ELECTRICITY AND MAGNETISM. Second Edition, revised and greatly enlarged. Fcap. 8vo. 5s. 6d.

THE MATHEMATICAL THEORY OF PERFECTLY ELASTIC SOLIDS, with a Short Account of Viscous Fluids. By W J. IBBETSON, late Senior Scholar of Clare College, Cambridge. 8vo. 21s.

NATURE'S STORY BOOKS. I. Sunshine. By AMY JOHNSON, LL.A. Illustrated. [*In the Press.*

EXAMPLES IN PHYSICS. With Answers and Solutions. By D. E. JONES, B.Sc., late Professor of Physics, University College of Wales, Aberystwith. Fcap. 8vo. 3s. 6d.

ELEMENTARY LESSONS IN HEAT, LIGHT, AND SOUND. By the Same. Globe 8vo. 2s. 6d.

CONTRIBUTIONS TO SOLAR PHYSICS. By J. NORMAN LOCKYER. F.R.S. With Illustrations. Royal 8vo. 31s. 6d.

MODERN VIEWS OF ELECTRICITY. By OLIVER J. LODGE, F.R.S., Professor of Physics, University College, Liverpool. Illustrated. Crown 8vo. 6s. 6d.

QUESTIONS AND EXAMPLES ON EXPERIMENTAL PHYSICS— Sound, Light, Heat, Electricity, and Magnetism. By B. LOEWY, Examiner in Experimental Physics to the College of Preceptors. Fcap. 8vo. 2s.

A GRADUATED COURSE OF NATURAL SCIENCE FOR ELEMENTARY AND TECHNICAL SCHOOLS AND COLLEGES. By the Same. In Three Parts. Part I. FIRST YEAR'S COURSE. Globe 8vo. 2s. Part II. 2s. 6d.

NUMERICAL TABLES AND CONSTANTS IN ELEMENTARY SCIENCE. By S. LUPTON, M.A. Extra fcap. 8vo. New Edition. 2s. 6d.

PHYSICAL ARITHMETIC. By A. MACFARLANE, D.Sc., late Examiner in Mathematics at the University of Edinburgh. Crown 8vo. 7s. 6d.

SOUND: A Series of Simple Experiments. By A. M. MAYER, Professor of Physics in the Stevens Institute of Technology. Illustrated. Crown 8vo. 3s. 6d.

LIGHT: A Series of Simple Experiments. By A. M. MAYER and C. BARNARD. Illustrated. Crown 8vo. 2s. 6d.

MACMILLAN & CO., LONDON.

Some of Messrs. Macmillan & Co.'s Books on Physics.

GLEANINGS IN SCIENCE: Popular Lectures. By Rev. GERALD MOLLOY, D.Sc., Rector of the Catholic University of Ireland. 8vo. 7s. 6d.

NEWTON.—PRINCIPIA. Edited by Professor Sir W. THOMSON, P.R.S., and Professor BLACKBURNE. 4to. 31s. 6d.

THE FIRST THREE SECTIONS OF NEWTON'S PRINCIPIA. With Notes, Illustrations, and Problems. By P. FROST, M.A., D.Sc. Third Edition. 8vo. 12s.

A TREATISE ON OPTICS. By S. PARKINSON, D.D., F.R.S., late Tutor of St. John's College, Cambridge. Fourth Edition. Crown 8vo. 10s. 6d.

THERMODYNAMICS OF THE STEAM-ENGINE AND OTHER HEAT-ENGINES. By CECIL H. PEABODY, Associate Professor of Steam Engineering, Massachusetts Institute of Technology. 8vo. 21s.

STEAM: An Elementary Treatise. By JOHN PERRY, Professor of Applied Mechanics, Technical College, Finsbury. 18mo. 4s. 6d.

ELEMENTS OF PHYSICAL MANIPULATION. By Professor EDWARD C. PICKERING. Medium 8vo. Part I., 12s. 6d. Part II., 14s.

THE THEORY OF LIGHT. By THOMAS PRESTON, M.A. Illustrated. 8vo. 15s. net.

THE THEORY OF HEAT. By the Same. 8vo. [*In preparation.*

THE THEORY OF SOUND. By Lord RAYLEIGH, F.R.S. 8vo. Vol. I., 12s. 6d. Vol II., 12s. 6d. [Vol. III. *In the Press.*

AN ELEMENTARY TREATISE ON HEAT, IN RELATION TO STEAM AND THE STEAM-ENGINE. By G. SHANN, M.A. Crown 8vo. 4s. 6d.

POLARISATION OF LIGHT. By the late W. SPOTTISWOODE, F.R.S Illustrated. Crown 8vo. 3s. 6d.

LESSONS IN ELEMENTARY PRACTICAL PHYSICS. By BALFOUR STEWART, F.R.S., and W. W. HALDANE GEE, B.Sc. Crown 8vo. Vol. I., GENERAL PHYSICAL PROCESSES. 6s. Vol. II., ELECTRICITY AND MAGNETISM. 7s. 6d. [Vol. III., OPTICS, HEAT, AND SOUND. *In the Press.*

PRACTICAL PHYSICS FOR SCHOOLS AND THE JUNIOR STUDENTS OF COLLEGES. By the Same Authors. Globe 8vo. Vol. I., ELECTRICITY AND MAGNETISM. 2s. 6d.
[Vol. II., OPTICS, HEAT, AND SOUND. *In the Press.*

ON LIGHT. Burnett Lectures. By Sir G. G. STOKES, F.R.S., Lucasian Professor of Mathematics in the University of Cambridge. I., ON THE NATURE OF LIGHT. II., ON LIGHT AS A MEANS OF INVESTIGATION. III., ON THE BENEFICIAL EFFECTS OF LIGHT. Crown 8vo. 7s. 6d.
*** The Second and Third Courses may be had separately. Crown 8vo. 2s. 6d. each.

AN ELEMENTARY TREATISE ON SOUND. By W. H. STONE. Illustrated. Fcap. 8vo. 3s. 6d.

HEAT. By P. G. TAIT, Professor of Natural Philosophy in the University of Edinburgh. Crown 8vo. 6s.

LECTURES ON SOME RECENT ADVANCES IN PHYSICAL SCIENCE. By the Same. Third Edition. Crown 8vo. 9s.

SOUND AND MUSIC. An Elementary Treatise on the Physical Constitution of Musical Sounds and Harmony, including the Chief Acoustical Discoveries of Professor Helmholtz. By SEDLEY TAYLOR, M.A. Extra crown 8vo. 8s. 6d.

A COLLECTION OF EXAMPLES ON HEAT AND ELECTRICITY. By H. H. TURNER, Fellow of Trinity College, Cambridge. Cr. 8vo. 2s. 6d.

LIGHT: A Course of Experimental Optics, chiefly with the Lantern. By LEWIS WRIGHT. Illustrated. Crown 8vo. 7s. 6d.

ELEMENTARY LESSONS IN ELECTRICITY AND MAGNETISM. By SILVANUS P. THOMPSON, Principal and Professor of Physics in the Technical College, Finsbury. Illustrated. Fcap. 8vo. 4s. 6d.

MACMILLAN & CO., LONDON.

October 1891

A Catalogue
OF
Educational Books
PUBLISHED BY
Macmillan & Co.
BEDFORD STREET, STRAND, LONDON

CONTENTS

	PAGE		PAGE
CLASSICS—		PHYSICS	30
ELEMENTARY CLASSICS . . .	2	ASTRONOMY	32
CLASSICAL SERIES	4	HISTORICAL	33
CLASSICAL LIBRARY ; Texts, Commentaries, Translations . .	6	**NATURAL SCIENCES—**	
GRAMMAR, COMPOSITION, AND PHILOLOGY	9	CHEMISTRY	33
ANTIQUITIES, ANCIENT HISTORY, AND PHILOSOPHY . . .	12	PHYSICAL GEOGRAPHY, GEOLOGY, AND MINERALOGY . .	35
		BIOLOGY	36
MODERN LANGUAGES AND LITERATURE—		MEDICINE	39
ENGLISH	14	**HUMAN SCIENCES—**	
FRENCH	18	MENTAL AND MORAL PHILOSOPHY	39
GERMAN	20	POLITICAL ECONOMY . .	41
MODERN GREEK	21	LAW AND POLITICS . .	42
ITALIAN	21	ANTHROPOLOGY . . .	43
SPANISH	21	EDUCATION	43
MATHEMATICS—		**TECHNICAL KNOWLEDGE—**	
ARITHMETIC	22	CIVIL AND MECHANICAL ENGINEERING	44
BOOK-KEEPING	23	MILITARY AND NAVAL SCIENCE	44
ALGEBRA	23	AGRICULTURE . . .	45
EUCLID AND PURE GEOMETRY .	24	DOMESTIC ECONOMY . .	46
GEOMETRICAL DRAWING . .	25	BOOK-KEEPING . . .	46
MENSURATION	25	COMMERCE	46
TRIGONOMETRY	25	GEOGRAPHY	47
ANALYTICAL GEOMETRY . .	26	HISTORY	47
PROBLEMS AND QUESTIONS IN MATHEMATICS	27	ART	50
HIGHER PURE MATHEMATICS .	27	DIVINITY	51
MECHANICS	28		

A

CLASSICS.

Elementary Classics; Classical Series; Classical Library, (1) Texts, (2) Translations; Grammar, Composition, and Philology; Antiquities, Ancient History, and Philosophy.

*ELEMENTARY CLASSICS.

18mo, Eighteenpence each.

The following contain Introductions, Notes, **and Vocabularies**, and in some cases **Exercises**.

ACCIDENCE, LATIN, AND EXERCISES ARRANGED FOR BEGINNERS.—By W. WELCH, M.A., and C. G. DUFFIELD, M.A.

AESCHYLUS.—PROMETHEUS VINCTUS. By Rev. H. M. STEPHENSON, M.A.

ARRIAN.—SELECTIONS. With Exercises. By Rev. JOHN BOND, M.A., and Rev. A. S. WALPOLE, M.A.

AULUS GELLIUS, STORIES FROM.—Adapted for Beginners. With Exercises. By Rev. G. H. NALL, M.A., Assistant Master at Westminster.

CÆSAR.—THE HELVETIAN WAR. Being Selections from Book I. of The Gallic War. Adapted for Beginners. With Exercises. By W. WELCH, M.A., and C. G. DUFFIELD, M.A.

THE INVASION OF BRITAIN. Being Selections from Books IV. and V. of The Gallic War. Adapted for Beginners. With Exercises. By W. WELCH, M.A., and C. G. DUFFIELD, M.A.

SCENES FROM BOOKS V. AND VI. By C. COLBECK, M.A.

THE GALLIC WAR. BOOK I. By Rev. A. S. WALPOLE, M.A.

BOOKS II. AND III. By the Rev. W. G. RUTHERFORD, M.A., LL.D.

BOOK IV. By CLEMENT BRYANS, M.A., Assistant Master at Dulwich College.

BOOK V. By C. COLBECK, M.A., Assistant Master at Harrow.

BOOK VI. By the same Editor.

BOOK VII. By Rev. J. BOND, M.A., and Rev. A. S. WALPOLE, M.A.

THE CIVIL WAR. BOOK I. By M. MONTGOMREY, M.A.

CICERO.—DE SENECTUTE. By E. S. SHUCKBURGH, M.A.

DE AMICITIA. By the same Editor.

STORIES OF ROMAN HISTORY. Adapted for Beginners. With Exercises. By Rev. G. E. JEANS, M.A., and A. V. JONES, M.A.

EURIPIDES.—ALCESTIS. By Rev. M. A. BAYFIELD, M.A.

MEDEA. By A. W. VERRALL, Litt.D., and Rev. M. A. BAYFIELD, M.A.
[*In the Press.*

HECUBA. By Rev. J. BOND, M.A., and Rev. A. S. WALPOLE, M.A.

EUTROPIUS.—Adapted for Beginners. With Exercises. By W. WELCH, M.A., and C. G. DUFFIELD, M.A.

HERODOTUS. TALES FROM HERODOTUS. Atticised by G. S. FARNELL, M.A.

HOMER.—ILIAD. BOOK I. By Rev. J. BOND, M.A., and Rev. A. S. WALPOLE, M.A.

BOOK XVIII. By S. R. JAMES, M.A., Assistant Master at Eton.

ODYSSEY. BOOK I. By Rev. J. BOND, M.A., and Rev. A. S. WALPOLE, M.A.

ELEMENTARY CLASSICS

HORACE.—ODES. BOOKS I.-IV. By T. E. PAGE, M.A., Assistant Master at the Charterhouse. Each 1s. 6d.
LIVY.—BOOK I. By H. M. STEPHENSON, M.A.
  BOOK XXI. Adapted from Mr. Capes's Edition. By J. E. MELHUISH, M.A.
  BOOK XXII. By the same.
  THE HANNIBALIAN WAR. Being part of the XXI. and XXII. BOOKS OF LIVY adapted for Beginners. By G. C. MACAULAY, M.A.
  THE SIEGE OF SYRACUSE. Being part of the XXIV. and XXV. BOOKS OF LIVY, adapted for Beginners. With Exercises. By G. RICHARDS, M.A., and Rev. A. S. WALPOLE, M.A.
  LEGENDS OF ANCIENT ROME. Adapted for Beginners. With Exercises. By H. WILKINSON, M.A.
LUCIAN.—EXTRACTS FROM LUCIAN. With Exercises. By Rev. J. BOND, M.A., and Rev. A. S. WALPOLE, M.A.
NEPOS.—SELECTIONS ILLUSTRATIVE OF GREEK AND ROMAN HISTORY. With Exercises. By G. S. FARNELL, M.A.
OVID.—SELECTIONS. By E. S. SHUCKBURGH, M.A.
  EASY SELECTIONS FROM OVID IN ELEGIAC VERSE. With Exercises. By H. WILKINSON, M.A.
  STORIES FROM THE METAMORPHOSES. With Exercises. By Rev. J. BOND, M.A., and Rev. A. S. WALPOLE, M.A.
PHÆDRUS.—SELECT FABLES. Adapted for Beginners. With Exercises. By Rev. A. S. WALPOLE, M.A.
THUCYDIDES.—THE RISE OF THE ATHENIAN EMPIRE. BOOK I. Chs. 89-117 and 228-238. With Exercises. By F. H. COLSON, M.A.
VIRGIL.—SELECTIONS. By E. S. SHUCKBURGH, M.A.
  BUCOLICS. By T. E. PAGE, M.A.
  GEORGICS. BOOK I. By the same Editor.
  BOOK II. By Rev. J. H. SKRINE, M.A.
  ÆNEID. BOOK I. By Rev. A. S. WALPOLE, M.A.
  BOOK II. By T. E. PAGE, M.A.
  BOOK III. By the same Editor.
  BOOK IV. By Rev. H. M. STEPHENSON, M.A.
  BOOK V. By Rev. A. CALVERT, M.A.
  BOOK VI. By T. E. PAGE, M.A.
  BOOK VII. By Rev. A. CALVERT, M.A.
  BOOK VIII. By the same Editor.
  BOOK IX. By Rev. H. M. STEPHENSON, M.A.
  BOOK X. By S. G. OWEN, M.A.
XENOPHON.—ANABASIS. Selections, adapted for Beginners. With Exercises. By W. WELCH, M.A., and C. G. DUFFIELD, M.A.
  BOOK I. With Exercises. By E. A. WELLS, M.A.
  BOOK I. By Rev. A. S. WALPOLE, M.A.
  BOOK II. By the same Editor.
  BOOK III. By Rev. G. H. NALL, M.A.
  BOOK IV. By Rev. E. D. STONE, M.A.
  SELECTIONS FROM BOOK IV. With Exercises. By the same Editor.
  SELECTIONS FROM THE CYROPÆDIA. With Exercises. By A. H. COOKE, M.A., Fellow and Lecturer of King's College, Cambridge.

The following contain Introductions and Notes, **but no Vocabulary** :—

CICERO.—SELECT LETTERS. By Rev. G. E. JEANS, M.A.
HERODOTUS.—SELECTIONS FROM BOOKS VII. AND VIII. THE EXPEDITION OF XERXES. By A. H. COOKE, M.A.
HORACE.—SELECTIONS FROM THE SATIRES AND EPISTLES. By Rev. W. J. V. BAKER, M.A.
  SELECT EPODES AND ARS POETICA. By H. A. DALTON, M.A., Assistant Master at Winchester.

PLATO.—EUTHYPHRO AND MENEXENUS. By C. E. Graves, M.A.
TERENCE.—SCENES FROM THE ANDRIA. By F. W. Cornish, M.A., Assistant Master at Eton.
THE GREEK ELEGIAC POETS.—FROM CALLINUS TO CALLIMACHUS. Selected by Rev. Herbert Kynaston, D.D.
THUCYDIDES.—BOOK IV. Chs. 1-41. THE CAPTURE OF SPHACTERIA. By C. E. Graves, M.A.

CLASSICAL SERIES FOR COLLEGES AND SCHOOLS.
Fcap. 8vo.

ÆSCHINES.—IN CTESIPHONTA. By Rev. T. Gwatkin, M.A., and E. S. Shuckburgh, M.A. 5s.
ÆSCHYLUS.—PERSÆ. By A. O. Prickard, M.A., Fellow and Tutor of New College, Oxford. With Map. 2s. 6d.
  SEVEN AGAINST THEBES. SCHOOL EDITION. By A. W. Verrall, Litt.D., Fellow of Trinity College, Cambridge, and M. A. Bayfield, M.A., Headmaster of Christ's College, Brecon. 2s. 6d.
ANDOCIDES.—DE MYSTERIIS. By W. J. Hickie, M.A. 2s. 6d.
ATTIC ORATORS.—Selections from ANTIPHON, ANDOCIDES, LYSIAS, ISOCRATES, and ISAEUS. By R. C. Jebb, Litt.D., Regius Professor of Greek in the University of Cambridge. 5s.
*CÆSAR.—THE GALLIC WAR. By Rev. John Bond, M.A., and Rev. A. S. Walpole, M.A. With Maps. 4s. 6d.
CATULLUS.—SELECT POEMS. Edited by F. P. Simpson, B.A. 3s. 6d. The Text of this Edition is carefully expurgated for School use.
*CICERO.—THE CATILINE ORATIONS. By A. S. Wilkins, Litt.D., Professor of Latin in the Owens College, Victoria University, Manchester. 2s. 6d.
  PRO LEGE MANILIA. By Prof. A. S. Wilkins, Litt.D. 2s. 6d.
  THE SECOND PHILIPPIC ORATION. By John E. B. Mayor, M.A., Professor of Latin in the University of Cambridge. 3s. 6d.
  PRO ROSCIO AMERINO. By E. H. Donkin, M.A. 2s. 6d.
  PRO P. SESTIO. By Rev. H. A. Holden, Litt.D. 3s. 6d.
  SELECT LETTERS. Edited by R. Y. Tyrrell, M.A. 4s. 6d.
DEMOSTHENES.—DE CORONA. By B. Drake, M.A. 7th Edition, revised by E. S. Shuckburgh, M.A. 3s. 6d.
  ADVERSUS LEPTINEM. By Rev. J. R. King, M.A., Fellow and Tutor of Oriel College, Oxford. 2s. 6d.
  THE FIRST PHILIPPIC. By Rev. T. Gwatkin, M.A. 2s. 6d.
  IN MIDIAM. By Prof. A. S. Wilkins, Litt.D., and Herman Hager, Ph.D., of the Owens College, Victoria University, Manchester. [In preparation.
EURIPIDES.—HIPPOLYTUS. By Rev. J. P. Mahaffy, D.D., Fellow of Trinity College, and Professor of Ancient History in the University of Dublin, and J. B. Bury, M.A., Fellow of Trinity College, Dublin. 2s. 6d.
  MEDEA. By A. W. Verrall, Litt.D., Fellow of Trinity College, Cambridge. 2s. 6d.
  IPHIGENIA IN TAURIS. By E. B. England, M.A. 3s.
  ION. By M. A. Bayfield, M.A., Headmaster of Christ's College, Brecon. 2s. 6d.
  BACCHAE. By R. Y. Tyrrell, M.A., Regius Professor of Greek in the University of Dublin. [In preparation.
HERODOTUS.—BOOK III. By G. C. Macaulay, M.A. 2s. 6d.
  BOOK V. By J. Strachan, M.A., Professor of Greek in the Owens College, Victoria University, Manchester. [In preparation.
  BOOK VI. By the same. 3s. 6d.
  BOOK VII. By Mrs. A. F. Butler. 3s. 6d.
HESIOD.—THE WORKS AND DAYS. By W. T. Lendrum M.A., Assistant Master at Dulwich College. [In preparation.

CLASSICAL SERIES

HOMER.—ILIAD. BOOKS I., IX., XI., XVI.-XXIV. THE STORY OF ACHILLES. By the late J. H. PRATT, M.A., and WALTER LEAF, Litt.D., Fellows of Trinity College, Cambridge. 5s.
 ODYSSEY. BOOK IX. By Prof. JOHN E. B. MAYOR. 2s. 6d.
 ODYSSEY. BOOKS XXI.-XXIV. THE TRIUMPH OF ODYSSEUS. By S. G. HAMILTON, B.A., Fellow of Hertford College, Oxford. 2s. 6d.
HORACE.—*THE ODES. By T. E. PAGE, M.A., Assistant Master at the Charterhouse. 5s. (BOOKS I., II., III., and IV. separately, 2s. each.)
 THE SATIRES. By ARTHUR PALMER, M.A., Professor of Latin in the University of Dublin. 5s.
 THE EPISTLES AND ARS POETICA. By A. S. WILKINS, Litt.D., Professor of Latin in the Owens College, Victoria University, Manchester. 5s.
ISAEOS.—THE ORATIONS. By WILLIAM RIDGEWAY, M.A., Professor of Greek in Queen's College, Cork. [*In preparation.*
JUVENAL.—*THIRTEEN SATIRES. By E. G. HARDY, M.A. 5s. The Text is carefully expurgated for School use.
 SELECT SATIRES. By Prof. JOHN E. B. MAYOR. X. and XI. 3s. 6d. XII.-XVI. 4s. 6d.
LIVY.—*BOOKS II. and III. By Rev. H. M. STEPHENSON, M.A. 3s. 6d.
 *BOOKS XXI. and XXII. By Rev. W. W. CAPES, M.A. With Maps. 4s. 6d.
 *BOOKS XXIII. and XXIV. By G. C. MACAULAY, M.A. With Maps. 3s. 6d.
 *THE LAST TWO KINGS OF MACEDON. EXTRACTS FROM THE FOURTH AND FIFTH DECADES OF LIVY. By F. H. RAWLINS, M.A., Assistant Master at Eton. With Maps. 2s. 6d.
 THE SUBJUGATION OF ITALY. SELECTIONS FROM THE FIRST DECADE. By G. E. MARINDIN, M.A. [*In preparation.*
LUCRETIUS.—BOOKS I.-III. By J. H. WARBURTON LEE, M.A., Assistant Master at Rossall. 3s. 6d.
LYSIAS.—SELECT ORATIONS. By E. S. SHUCKBURGH, M.A. 5s.
MARTIAL.—SELECT EPIGRAMS. By Rev. H. M. STEPHENSON, M.A. 5s.
*OVID.—FASTI. By G. H. HALLAM, M.A., Assistant Master at Harrow. With Maps. 3s. 6d.
 *HEROIDUM EPISTULÆ XIII. By E. S. SHUCKBURGH, M.A. 3s. 6d.
 METAMORPHOSES. BOOKS I.-III. By C. SIMMONS, M.A. [*In preparation.*
 BOOKS XIII. and XIV. By the same Editor. 3s. 6d.
PLATO.—LACHES. By M. T. TATHAM, M.A. 2s. 6d.
 THE REPUBLIC. BOOKS I.-V. By T. H. WARREN, M.A., President of Magdalen College, Oxford. 5s.
PLAUTUS.—MILES GLORIOSUS. By R. Y. TYRRELL, M.A., Regius Professor of Greek in the University of Dublin. 2d Ed., revised. 3s. 6d.
 AMPHITRUO. By ARTHUR PALMER, M.A., Professor of Latin in the University of Dublin. 3s. 6d.
 CAPTIVI. By A. R. S. HALLIDIE, M.A. 3s. 6d.
PLINY.—LETTERS. BOOKS I. and II. By J. COWAN, M.A., Assistant Master at the Manchester Grammar School. 3s.
 LETTERS. BOOK III. By Prof. JOHN E. B. MAYOR. With Life of Pliny by G. H. RENDALL, M.A. 3s. 6d.
PLUTARCH.—LIFE OF THEMISTOKLES. By Rev. H. A. HOLDEN, Litt.D. 3s. 6d.
 LIVES OF GALBA AND OTHO. By E. G. HARDY, M.A. 5s.
POLYBIUS.—THE HISTORY OF THE ACHÆAN LEAGUE AS CONTAINED IN THE REMAINS OF POLYBIUS. By W. W. CAPES, M.A. 5s.
PROPERTIUS.—SELECT POEMS. By Prof. J. P. POSTGATE, Litt.D., Fellow of Trinity College, Cambridge. 2d Ed., revised. 5s.
SALLUST.—*CATILINA and JUGURTHA. By C. MERIVALE, D.D., Dean of Ely. 3s. 6d. Or separately, 2s. each.
 *BELLUM CATULINÆ. By A. M. COOK, M.A., Assistant Master at St. Paul's School. 2s. 6d.
 JUGURTHA. By the same Editor. [*In preparation.*

TACITUS.—THE ANNALS. BOOKS I. and II. By J. S. REID, Litt.D. [*In prep.*
THE ANNALS. BOOK VI. By A. J. CHURCH, M.A., and W. J. BRODRIBB, M.A. 2s.
THE HISTORIES. BOOKS I. and II. By A. D. GODLEY, M.A., Fellow of Magdalen College, Oxford. 3s. 6d. BOOKS III.-V. By the same. 3s. 6d.
AGRICOLA and **GERMANIA.** By A. J. CHURCH, M.A., and W. J. BRODRIBB, M.A. 3s. 6d. Or separately, 2s. each.
TERENCE.—HAUTON TIMORUMENOS. By E. S. SHUCKBURGH, M.A. 2s. 6d. With Translation. 3s. 6d.
PHORMIO. By Rev. JOHN BOND, M.A., and Rev. A. S. WALPOLE, M.A. 2s. 6d.
THUCYDIDES.—BOOK I. By C. BRYANS, M.A. [*In preparation.*
BOOK II. By E. C. MARCHANT, M.A., Assistant Master at St. Paul's. 3s. 6d.
BOOK III. By C. BRYANS, M.A. [*In preparation.*
BOOK IV. By C. E. GRAVES, M.A., Classical Lecturer at St. John's College, Cambridge. 3s. 6d.
BOOK V. By the same Editor. 3s. 6d.
BOOKS VI. AND VII. THE SICILIAN EXPEDITION. By Rev. PERCIVAL FROST, M.A. With Map. 3s. 6d.
BOOK VIII. By Prof. T. G. TUCKER, Litt.D. [*In the Press.*
TIBULLUS.—SELECT POEMS. By Prof. J. P. POSTGATE, Litt.D. [*In preparation.*
VIRGIL.—ÆNEID. BOOKS II. AND III. THE NARRATIVE OF ÆNEAS. By E. W. HOWSON, M.A., Assistant Master at Harrow. 2s.
XENOPHON.—*THE ANABASIS. BOOKS I.-IV. By Profs. W. W. GOODWIN and J. W. WHITE. Adapted to Goodwin's Greek Grammar. With Map. 3s. 6d.
HELLENICA. BOOKS I. AND II. By H. HAILSTONE, B.A. With Map. 2s. 6d.
CYROPÆDIA. BOOKS VII. AND VIII. By A. GOODWIN, M.A., Professor of Classics in University College, London. 2s. 6d.
MEMORABILIA SOCRATIS. By A. R. CLUER, B.A., Balliol College, Oxford. 5s.
HIERO. By Rev. H. A. HOLDEN, Litt.D., LL.D. 2s. 6d.
OECONOMICUS. By the same. With Lexicon. 5s.

CLASSICAL LIBRARY.

Texts, Edited with **Introductions and Notes**, for the use of Advanced Students; **Commentaries and Translations.**

ÆSCHYLUS.—THE SUPPLICES. A Revised Text, with Translation. By T. G. TUCKER, Litt.D., Professor of Classical Philology in the University of Melbourne. 8vo. 10s. 6d.
THE SEVEN AGAINST THEBES. With Translation. By A. W. VERRALL, Litt.D., Fellow of Trinity College, Cambridge. 8vo. 7s. 6d.
AGAMEMNON. With Translation. By A. W. VERRALL, Litt.D. 8vo. 12s.
AGAMEMNON, CHOEPHORŒ, AND EUMENIDES. By A. O. PRICKARD, M.A., Fellow and Tutor of New College, Oxford. 8vo. [*In preparation.*
THE EUMENIDES. With Verse Translation. By BERNARD DRAKE, M.A. 8vo. 5s.
ANTONINUS, MARCUS AURELIUS.—BOOK IV. OF THE MEDITATIONS. With Translation. By HASTINGS CROSSLEY, M.A. 8vo. 6s.
ARISTOTLE.—THE METAPHYSICS. BOOK I. Translated by a Cambridge Graduate. 8vo. 5s.
THE POLITICS. By R. D. HICKS, M.A., Fellow of Trinity College, Cambridge. 8vo. [*In the Press.*
THE POLITICS. Translated by Rev. J. E. C. WELLDON, M.A., Headmaster of Harrow. Cr. 8vo. 10s. 6d.
THE RHETORIC. Translated by the same. Cr. 8vo. 7s. 6d.
AN INTRODUCTION TO ARISTOTLE'S RHETORIC. With Analysis, Notes, and Appendices. By E. M. COPE, Fellow and late Tutor of Trinity College, Cambridge. 8vo. 14s.
THE ETHICS. Translated by Rev. J. E. C. WELLDON, M.A. Cr. 8vo. [*In prep.*

CLASSICAL LIBRARY

THE SOPHISTICI ELENCHI. With Translation. By E. POSTE, M.A., Fellow of Oriel College, Oxford. 8vo. 8s. 6d.
ON THE CONSTITUTION OF ATHENS. Edited by J. E. SANDYS, Litt.D.
ON THE CONSTITUTION OF ATHENS. Translated by E. POSTE, M.A. Cr. 8vo. 3s. 6d.
ON THE ART OF POETRY. A Lecture. By A. O. PRICKARD, M.A., Fellow and Tutor of New College, Oxford. Cr. 8vo. 3s. 6d.
ARISTOPHANES.—THE BIRDS. Translated into English Verse. By B. H. KENNEDY, D.D. Cr. 8vo. 6s. Help Notes to the Same, for the Use of Students. 1s. 6d.
ATTIC ORATORS.—FROM ANTIPHON TO ISAEOS. By R. C. JEBB, Litt.D., Regius Professor of Greek in the University of Cambridge. 2 vols. 8vo. 25s.
BABRIUS.—With Lexicon. By Rev. W. G. RUTHERFORD, M.A., LL.D., Headmaster of Westminster. 8vo. 12s. 6d.
CICERO.—THE ACADEMICA. By J. S. REID, Litt.D., Fellow of Caius College, Cambridge. 8vo. 15s.
THE ACADEMICS. Translated by the same. 8vo. 5s. 6d.
SELECT LETTERS. After the Edition of ALBERT WATSON, M.A. Translated by G. E. JEANS, M.A., Fellow of Hertford College, Oxford. Cr. 8vo. 10s. 6d.
EURIPIDES.—MEDEA. Edited by A. W. VERRALL, Litt.D. 8vo. 7s. 6d.
IPHIGENEIA AT AULIS. Edited by E. B. ENGLAND, M.A. 8vo. 7s. 6d.
*INTRODUCTION TO THE STUDY OF EURIPIDES. By Professor J. P. MAHAFFY. Fcap. 8vo. 1s. 6d. (*Classical Writers.*)
HERODOTUS.—BOOKS I.-III. THE ANCIENT EMPIRES OF THE EAST. Edited by A. H. SAYCE, Deputy-Professor of Comparative Philology, Oxford. 8vo. 16s.
BOOKS IV.-IX. Edited by R. W. MACAN, M.A., Reader in Ancient History in the University of Oxford. 8vo. [*In preparation.*
THE HISTORY. Translated by G. C. MACAULAY, M.A. 2 vols. Cr. 8vo. 18s.
HERONDAS.—ΗΡΩΝΔΟΥ ΜΙΜΙΑΜΒΟΙ. A First Recension by Rev. W. G. RUTHERFORD, M.A., LL.D., Headmaster of Westminster. 8vo, sewed. 2s. net.
HOMER.—THE ILIAD. By WALTER LEAF, Litt.D. 8vo. Books I.-XII. 14s. Books XIII.-XXIV. 14s.
THE ILIAD. Translated into English Prose by ANDREW LANG, M.A., WALTER LEAF, Litt.D., and ERNEST MYERS, M.A. Cr. 8vo. 12s. 6d.
THE ODYSSEY. Done into English by S. H. BUTCHER, M.A., Professor of Greek in the University of Edinburgh, and ANDREW LANG, M.A. Cr. 8vo. 6s.
*INTRODUCTION TO THE STUDY OF HOMER. By the Right Hon. W. E. GLADSTONE. 18mo. 1s. (*Literature Primers.*)
HOMERIC DICTIONARY. Translated from the German of Dr. G. AUTENRIETH by R. P. KEEP, Ph.D. Illustrated. Cr. 8vo. 6s.
HORACE.—Translated by J. LONSDALE, M.A., and S. LEE, M.A. Gl. 8vo. 3s. 6d.
STUDIES, LITERARY AND HISTORICAL, IN THE ODES OF HORACE. By A. W. VERRALL, Litt.D. 8vo. 8s. 6d.
JUVENAL.—THIRTEEN SATIRES OF JUVENAL. By JOHN E. B. MAYOR, M.A., Professor of Latin in the University of Cambridge. Cr. 8vo. 2 vols. 10s. 6d. each. Vol. I. 10s. 6d. Vol. II. 10s. 6d.
THIRTEEN SATIRES. Translated by ALEX. LEEPER, M.A., LL.D., Warden of Trinity College, Melbourne. Cr. 8vo. 3s. 6d.
KTESIAS.—THE FRAGMENTS OF THE PERSIKA OF KTESIAS. By JOHN GILMORE, M.A. 8vo. 8s. 6d.
LIVY.—BOOKS I.-IV. Translated by Rev. H. M. STEPHENSON, M.A. [*In prep.*
BOOKS XXI.-XXV. Translated by A. J. CHURCH, M.A., and W. J. BRODRIBB, M.A. Cr. 8vo. 7s. 6d.
*INTRODUCTION TO THE STUDY OF LIVY. By Rev. W. W. CAPES, M.A. Fcap. 8vo. 1s. 6d. (*Classical Writers.*)
LONGINUS.—ON THE SUBLIME. Translated by H. L. HAVELL, B.A. With Introduction by ANDREW LANG. Cr. 8vo. 4s. 6d.

MARTIAL.—BOOKS I. AND II. OF THE EPIGRAMS. By Prof. JOHN E. B. MAYOR, M.A. 8vo. [*In the Press.*

MELEAGER.—FIFTY POEMS OF MELEAGER. Translated by WALTER HEADLAM. Fcap. 4to. 7s. 6d.

PAUSANIAS.—DESCRIPTION OF GREECE. Translated with Commentary by J. G. FRAZER, M.A., Fellow of Trinity College, Cambridge. [*In prep.*

PHRYNICHUS.—THE NEW PHRYNICHUS; being a Revised Text of the Ecloga of the Grammarian Phrynichus. With Introduction and Commentary by Rev. W. G. RUTHERFORD, M.A., LL.D., Headmaster of Westminster. 8vo. 18s.

PINDAR.—THE EXTANT ODES OF PINDAR. Translated by ERNEST MYERS, M.A. Cr. 8vo. 5s.

THE OLYMPIAN AND PYTHIAN ODES. Edited, with an Introductory Essay, by BASIL GILDERSLEEVE, Professor of Greek in the Johns Hopkins University, U.S.A. Cr. 8vo. 7s. 6d.

THE NEMEAN ODES. By J. B. BURY, M.A., Fellow of Trinity College, Dublin. 8vo. 12s.

THE ISTHMIAN ODES. By the same Editor. [*In the Press.*

PLATO.—PHÆDO. By R. D. ARCHER-HIND, M.A., Fellow of Trinity College, Cambridge. 8vo. 8s. 6d.

PHÆDO. By W. D. GEDDES, LL.D., Principal of the University of Aberdeen. 8vo. 8s. 6d.

TIMAEUS. With Translation. By R. D. ARCHER-HIND, M.A. 8vo. 16s.

THE REPUBLIC OF PLATO. Translated by J. LL. DAVIES, M.A., and D. J. VAUGHAN, M.A. 18mo. 4s. 6d.

EUTHYPHRO, APOLOGY, CRITO, AND PHÆDO. Translated by F. J. CHURCH. 18mo. 4s. 6d.

PHÆDRUS, LYSIS, AND PROTAGORAS. Translated by J. WRIGHT, M.A. 18mo. 4s. 6d.

PLAUTUS.—THE MOSTELLARIA. By WILLIAM RAMSAY, M.A. Edited by G. G. RAMSAY, M.A., Professor of Humanity in the University of Glasgow. 8vo. 14s.

PLINY.—CORRESPONDENCE WITH TRAJAN. C. Plinii Caecilii Secundi Epistulæ ad Traianum Imperatorem cum Eiusdem Responsis. By E. G. HARDY, M.A. 8vo. 10s. 6d.

POLYBIUS.—THE HISTORIES OF POLYBIUS. Translated by E. S. SHUCKBURGH, M.A. 2 vols. Cr. 8vo. 24s.

SALLUST.—CATILINE AND JUGURTHA. Translated by A. W. POLLARD, B.A. Cr. 8vo. 6s. THE CATILINE (separately). 3s.

SOPHOCLES.—ŒDIPUS THE KING. Translated into English Verse by E. D. A. MORSHEAD, M.A., Assistant Master at Winchester. Fcap. 8vo. 3s. 6d.

TACITUS.—THE ANNALS. By G. O. HOLBROOKE, M.A., Professor of Latin in Trinity College, Hartford, U.S.A. With Maps. 8vo. 16s.

THE ANNALS. Translated by A. J. CHURCH, M.A., and W. J. BRODRIBB, M.A. With Maps. Cr. 8vo. 7s. 6d.

THE HISTORIES. By Rev. W. A. SPOONER, M.A., Fellow and Tutor of New College, Oxford. 8vo. 16s.

THE HISTORY. Translated by A. J. CHURCH, M.A., and W. J. BRODRIBB, M.A. With Map. Cr. 8vo. 6s.

THE AGRICOLA AND GERMANY, WITH THE DIALOGUE ON ORATORY. Translated by A. J. CHURCH, M.A., and W. J. BRODRIBB, M.A. With Maps. Cr. 8vo. 4s. 6d.

*INTRODUCTION TO THE STUDY OF TACITUS. By A. J. CHURCH, M.A., and W. J. BRODRIBB, M.A. Fcap. 8vo. 1s. 6d. (*Classical Writers.*)

THEOCRITUS, BION, AND MOSCHUS. Translated by A. LANG, M.A. 18mo. 4s. 6d. Also an Edition on Large Paper. Cr. 8vo. 9s.

THUCYDIDES.—BOOK IV. A Revision of the Text, Illustrating the Principal Causes of Corruption in the Manuscripts of this Author. By Rev. W. G. RUTHERFORD, M.A., LL.D., Headmaster of Westminster. 8vo. 7s. 6d.

GRAMMAR, COMPOSITION, AND PHILOLOGY

BOOK VIII. By H. C. GOODHART, M.A., Fellow of Trinity College, Cambridge.
[*In the Press.*
VIRGIL.—Translated by J. LONSDALE, M.A., and S. LEE, M.A. Gl. 8vo. 3s. 6d.
THE ÆNEID. Translated by J. W. MACKAIL, M.A., Fellow of Balliol College, Oxford. Cr. 8vo. 7s. 6d.
XENOPHON.—Translated by H. G. DAKYNS, M.A. In four vols. Cr. 8vo. Vol. I., containing "The Anabasis" and Books I. and II. of "The Hellenica." 10s. 6d. Vol. II. "Hellenica" III.-VII., and the two Polities—"Athenian" and "Laconian," the "Agesilaus," and the tract on "Revenues." With Maps and Plans.
[*In the Press.*

GRAMMAR, COMPOSITION, & PHILOLOGY.

*BELCHER.—SHORT EXERCISES IN LATIN PROSE COMPOSITION AND EXAMINATION PAPERS IN LATIN GRAMMAR. Part I. By Rev. H. BELCHER, LL.D., Rector of the High School, Dunedin, N.Z. 18mo. 1s. 6d.
KEY, for Teachers only. 18mo. 3s. 6d.
*Part II., On the Syntax of Sentences, with an Appendix, including EXERCISES IN LATIN IDIOMS, etc. 18mo. 2s. KEY, for Teachers only. 18mo. 3s.
BLACKIE.—GREEK AND ENGLISH DIALOGUES FOR USE IN SCHOOLS AND COLLEGES. By JOHN STUART BLACKIE, Emeritus Professor of Greek in the University of Edinburgh. New Edition. Fcap. 8vo. 2s. 6d.
A GREEK PRIMER, COLLOQUIAL AND CONSTRUCTIVE. Cr. 8vo. 2s. 6d.
*BRYANS.—LATIN PROSE EXERCISES BASED UPON CÆSAR'S GALLIC WAR. With a Classification of Cæsar's Chief Phrases and Grammatical Notes on Cæsar's Usages. By CLEMENT BRYANS, M.A., Assistant Master at Dulwich College. Ex. fcap. 8vo. 2s. 6d. KEY, for Teachers only. 4s. 6d.
GREEK PROSE EXERCISES based upon Thucydides. By the same.
[*In preparation.*
COOKSON.—A LATIN SYNTAX. By CHRISTOPHER COOKSON, M.A., Assistant Master at St. Paul's School. 8vo. [*In preparation.*
CORNELL UNIVERSITY STUDIES IN CLASSICAL PHILOLOGY. Edited by I. FLAGG, W. G. HALE, and B. I. WHEELER. I. The *CUM*-Constructions: their History and Functions. By W. G. HALE. Part 1. Critical. 1s. 8d. net. Part 2. Constructive. 3s. 4d. net. II. Analogy and the Scope of its Application in Language. By B. I. WHEELER. 1s. 3d. net.
*EICKE.—FIRST LESSONS IN LATIN. By K. M. EICKE, B.A., Assistant Master at Oundle School. Gl. 8vo. 2s. 6d.
*ENGLAND.—EXERCISES ON LATIN SYNTAX AND IDIOM. ARRANGED WITH REFERENCE TO ROBY'S SCHOOL LATIN GRAMMAR. By E. B. ENGLAND, Assistant Lecturer at the Owens College, Victoria University, Manchester. Cr. 8vo. 2s. 6d. KEY, for Teachers only. 2s. 6d.
GILES.—A SHORT MANUAL OF PHILOLOGY FOR CLASSICAL STUDENTS. By P. GILES, M.A., Reader in Comparative Philology in the University of Cambridge. Cr. 8vo. [*In the Press.*
GOODWIN.—Works by W. W. GOODWIN, LL.D., D.C.L., Professor of Greek in Harvard University, U.S.A.
SYNTAX OF THE MOODS AND TENSES OF THE GREEK VERB. New Ed., revised and enlarged. 8vo. 14s.
*A GREEK GRAMMAR. Cr. 8vo. 6s.
*A GREEK GRAMMAR FOR SCHOOLS. Cr. 8vo. 3s. 6d.
GREENWOOD.—THE ELEMENTS OF GREEK GRAMMAR. Adapted to the System of Crude Forms. By J. G. GREENWOOD, sometime Principal of the Owens College, Manchester. Cr. 8vo. 5s. 6d.
HADLEY.—ESSAYS, PHILOLOGICAL AND CRITICAL. By JAMES HADLEY, late Professor in Yale College. 8vo. 16s.
HADLEY and ALLEN.—A GREEK GRAMMAR FOR SCHOOLS AND COLLEGES. By JAMES HADLEY, late Professor in Yale College. Revised and in part rewritten by F. DE F. ALLEN, Professor in Harvard College. Cr. 8vo. 6s.

HODGSON.—MYTHOLOGY FOR LATIN VERSIFICATION. A brief sketch of the Fables of the Ancients, prepared to be rendered into Latin Verse for Schools. By F. HODGSON, B.D., late Provost of Eton. New Ed., revised by F. C. HODGSON, M.A. 18mo. 3s.

*JACKSON.—FIRST STEPS TO GREEK PROSE COMPOSITION. By BLOMFIELD JACKSON, M.A., Assistant Master at King's College School. 18mo. 1s. 6d. KEY, for Teachers only. 18mo. 3s. 6d.

*SECOND STEPS TO GREEK PROSE COMPOSITION, with Miscellaneous Idioms, Aids to Accentuation, and Examination Papers in Greek Scholarship. By the same. 18mo. 2s. 6d. KEY, for Teachers only. 18mo. 3s. 6d.

KYNASTON.—EXERCISES IN THE COMPOSITION OF GREEK IAMBIC VERSE by Translations from English Dramatists. By Rev. H. KYNASTON, D.D., Professor of Classics in the University of Durham. With Vocabulary. Ex. fcap. 8vo. 5s.

KEY, for Teachers only. Ex. fcap. 8vo. 4s. 6d.

LUPTON.—*AN INTRODUCTION TO LATIN ELEGIAC VERSE COMPOSITION. By J. H. LUPTON, Sur-Master of St. Paul's School. Gl. 8vo. 2s. 6d. KEY TO PART II. (XXV.-C.) Gl. 8vo. 3s. 6d.

*AN INTRODUCTION TO LATIN LYRIC VERSE COMPOSITION. By the same. Gl. 8vo. 3s. KEY, for Teachers only. Gl. 8vo. 4s. 6d.

MACKIE.—PARALLEL PASSAGES FOR TRANSLATION INTO GREEK AND ENGLISH. With Indexes. By Rev. ELLIS C. MACKIE, M.A., Classical Master at Heversham Grammar School. Gl. 8vo. 4s. 6d.

*MACMILLAN.—FIRST LATIN GRAMMAR. By M. C. MACMILLAN, M.A. Fcap. 8vo. 1s. 6d.

MACMILLAN'S GREEK COURSE.—Edited by Rev. W. G. RUTHERFORD, M.A., LL.D., Headmaster of Westminster. Gl. 8vo.

*FIRST GREEK GRAMMAR—ACCIDENCE. By the Editor. 2s.
*FIRST GREEK GRAMMAR—SYNTAX. By the same. 2s.
ACCIDENCE AND SYNTAX. In one volume. 3s. 6d.
*EASY EXERCISES IN GREEK ACCIDENCE. By H. G. UNDERHILL, M.A., Assistant Master at St. Paul's Preparatory School. 2s.
*A SECOND GREEK EXERCISE BOOK. By Rev. W. A. HEARD, M.A., Headmaster of Fettes College, Edinburgh. 2s. 6d.
EASY EXERCISES IN GREEK SYNTAX. By Rev. G. H. NALL, M.A., Assistant Master at Westminster School. [In preparation.
MANUAL OF GREEK ACCIDENCE. By the Editor. [In preparation.
MANUAL OF GREEK SYNTAX. By the Editor. [In preparation.
ELEMENTARY GREEK COMPOSITION. By the Editor. [In preparation.

*MACMILLAN'S GREEK READER.—STORIES AND LEGENDS. A First Greek Reader, with Notes, Vocabulary, and Exercises. By F. H. COLSON, M.A., Headmaster of Plymouth College. Gl. 8vo. 3s.

MACMILLAN'S LATIN COURSE.—By A. M. COOK, M.A., Assistant Master at St. Paul's School.

*FIRST PART. Gl. 8vo. 3s. 6d.
*SECOND PART. 2s. 6d. [Third Part in preparation.

*MACMILLAN'S SHORTER LATIN COURSE.—By A. M. COOK, M.A. Being an abridgment of "Macmillan's Latin Course," First Part. Gl. 8vo. 1s. 6d. KEY. [In the Press.

*MACMILLAN'S LATIN READER.—A LATIN READER FOR THE LOWER FORMS IN SCHOOLS. By H. J. HARDY, M.A., Assistant Master at Winchester. Gl. 8vo. 2s. 6d.

*MARSHALL.—A TABLE OF IRREGULAR GREEK VERBS, classified according to the arrangement of Curtius's Greek Grammar. By J. M. MARSHALL, M.A., Headmaster of the Grammar School, Durham. 8vo. 1s.

MAYOR.—FIRST GREEK READER. By Prof. JOHN E. B. MAYOR, M.A., Fellow of St. John's College, Cambridge. Fcap. 8vo. 4s. 6d.

GRAMMAR, COMPOSITION, AND PHILOLOGY

MAYOR.—GREEK FOR BEGINNERS. By Rev. J. B. MAYOR, M.A., late Professor of Classical Literature in King's College, London. Part I., with Vocabulary, 1s. 6d. Parts II. and III., with Vocabulary and Index. Fcap. 8vo. 3s. 6d. Complete in one Vol. 4s. 6d.

NIXON.—PARALLEL EXTRACTS, Arranged for Translation into English and Latin, with Notes on Idioms. By J. E. NIXON, M.A., Fellow and Classical Lecturer, King's College, Cambridge. Part I.—Historical and Epistolary. Cr. 8vo. 3s. 6d.

PROSE EXTRACTS, Arranged for Translation into English and Latin, with General and Special Prefaces on Style and Idiom. By the same. I. Oratorical. II. Historical. III. Philosophical. IV. Anecdotes and Letters. 2d Ed., enlarged to 280 pp. Cr. 8vo. 4s. 6d. SELECTIONS FROM THE SAME. 3s. Translations of about 70 Extracts can be supplied to Schoolmasters (2s. 6d.), on application to the Author: and about 40 similarly of "Parallel Extracts." 1s. 6d. post free.

*PANTIN.—A FIRST LATIN VERSE BOOK. By W. E. P. PANTIN, M.A., Assistant Master at St. Paul's School. Gl. 8vo. 1s. 6d.

*PEILE.—A PRIMER OF PHILOLOGY. By J. PEILE, Litt.D., Master of Christ's College, Cambridge. 18mo. 1s.

*POSTGATE.—SERMO LATINUS. A short Guide to Latin Prose Composition. By Prof. J. P. POSTGATE, Litt.D., Fellow of Trinity College, Cambridge. Gl. 8vo. 2s. 6d. KEY to "Selected Passages." Gl. 8vo. 3s. 6d.

POSTGATE and VINCE.—A DICTIONARY OF LATIN ETYMOLOGY. By J. P. POSTGATE and C. A. VINCE. [In preparation.

POTTS.—*HINTS TOWARDS LATIN PROSE COMPOSITION. By A. W. POTTS, M.A., LL.D., late Fellow of St. John's College, Cambridge. Ex. fcap. 8vo. 3s.

*PASSAGES FOR TRANSLATION INTO LATIN PROSE. Edited with Notes and References to the above. Ex. fcap. 8vo. 2s. 6d. KEY, for Teachers only. 2s. 6d.

*PRESTON.—EXERCISES IN LATIN VERSE OF VARIOUS KINDS. By Rev. G. PRESTON. Gl. 8vo. 2s. 6d. KEY, for Teachers only. Gl. 8vo. 5s.

REID.—A GRAMMAR OF TACITUS. By J. S. REID, Litt.D., Fellow of Caius College, Cambridge. [In the Press.

A GRAMMAR OF VIRGIL. By the same. [In preparation.

ROBY.—Works by H. J. ROBY, M.A., late Fellow of St. John's College, Cambridge. A GRAMMAR OF THE LATIN LANGUAGE, from Plautus to Suetonius. Part I. Sounds, Inflexions, Word-formation, Appendices. Cr. 8vo. 9s. Part II. Syntax, Prepositions, etc. 10s. 6d.

*SCHOOL LATIN GRAMMAR. Cr. 8vo. 5s.

AN ELEMENTARY LATIN GRAMMAR. [In the Press.

*RUSH.—SYNTHETIC LATIN DELECTUS. With Notes and Vocabulary. By E. RUSH, B.A. Ex. fcap. 8vo. 2s. 6d.

*RUST.—FIRST STEPS TO LATIN PROSE COMPOSITION. By Rev. G. RUST, M.A. 18mo. 1s. 6d. KEY, for Teachers only. By W. M. YATES. 18mo. 3s. 6d.

RUTHERFORD.—Works by the Rev. W. G. RUTHERFORD, M.A., LL.D., Headmaster of Westminster.

REX LEX. A Short Digest of the principal Relations between the Latin, Greek, and Anglo-Saxon Sounds. 8vo. [In preparation.

THE NEW PHRYNICHUS; being a Revised Text of the Ecloga of the Grammarian Phrynichus. With Introduction and Commentary. 8vo. 18s. (See also *Macmillan's Greek Course*.)

SCHOLIA ARISTOPHANICA; being such Comments and Adscripts to the text of Aristophanes as are preserved in the Codex Ravennas, arranged, emended, and translated. 8vo. [In the Press.

SHUCKBURGH.—PASSAGES FROM LATIN AUTHORS FOR TRANSLATION INTO ENGLISH. Selected with a view to the needs of Candidates for the Cambridge Local, and Public Schools' Examinations. By E. S. SHUCKBURGH, M.A. Cr. 8vo. 2s.

*SIMPSON. — LATIN PROSE AFTER THE BEST AUTHORS: Cæsarian Prose. By F. P. SIMPSON, B.A. Ex. fcap. 8vo. 2s. 6d. KEY, for Teachers only. Ex. fcap. 8vo. 5s.

STRACHAN and WILKINS.—ANALECTA. Selected Passages for Translation. By J. S. STRACHAN, M.A., Professor of Greek, and A. S. WILKINS, Litt.D., Professor of Latin in the Owens College, Manchester. Cr. 8vo. 5s. KEY to Latin Passages. Cr. 8vo. Sewed, 6d. KEY to Greek Passages. Sewed, 6d.

THRING.—Works by the Rev. E. THRING, M.A., late Headmaster of Uppingham.
A LATIN GRADUAL. A First Latin Construing Book for Beginners. With Coloured Sentence Maps. Fcap. 8vo. 2s. 6d.
A MANUAL OF MOOD CONSTRUCTIONS. Fcap. 8vo. 1s. 6d.

*****WELCH and DUFFIELD.**—LATIN ACCIDENCE AND EXERCISES ARRANGED FOR BEGINNERS. By W. WELCH and C. G. DUFFIELD, Assistant Masters at Cranleigh School. 18mo. 1s. 6d.

WHITE.—FIRST LESSONS IN GREEK. Adapted to GOODWIN'S GREEK GRAMMAR, and designed as an introduction to the ANABASIS OF XENOPHON. By JOHN WILLIAMS WHITE, Assistant Professor of Greek in Harvard University, U.S.A. Cr. 8vo. 3s. 6d.

WRIGHT.—Works by J. WRIGHT, M.A., late Headmaster of Sutton Coldfield School.
A HELP TO LATIN GRAMMAR; or, the Form and Use of Words in Latin, with Progressive Exercises. Cr. 8vo. 4s. 6d.
THE SEVEN KINGS OF ROME. An Easy Narrative, abridged from the First Book of Livy by the omission of Difficult Passages; being a First Latin Reading Book, with Grammatical Notes and Vocabulary. Fcap. 8vo. 3s. 6d.
FIRST LATIN STEPS; OR, AN INTRODUCTION BY A SERIES OF EXAMPLES TO THE STUDY OF THE LATIN LANGUAGE. Cr. 8vo. 3s.
ATTIC PRIMER. Arranged for the Use of Beginners. Ex. fcap. 8vo. 2s. 6d.
A COMPLETE LATIN COURSE, comprising Rules with Examples, Exercises, both Latin and English, on each Rule, and Vocabularies. Cr. 8vo. 2s. 6d.

ANTIQUITIES, ANCIENT HISTORY, AND PHILOSOPHY.

ARNOLD.—A HISTORY OF THE EARLY ROMAN EMPIRE. By W. T. ARNOLD, M.A. [*In preparation.*

ARNOLD.—THE SECOND PUNIC WAR. Being Chapters from THE HISTORY OF ROME by the late THOMAS ARNOLD, D.D., Headmaster of Rugby. Edited, with Notes, by W. T. ARNOLD, M.A. With 8 Maps. Cr. 8vo. 5s.

*****BEESLY.**—STORIES FROM THE HISTORY OF ROME. By Mrs. BEESLY. Fcap. 8vo. 2s. 6d.

BLACKIE.—HORÆ HELLENICÆ. By JOHN STUART BLACKIE, Emeritus Professor of Greek in the University of Edinburgh. 8vo. 12s.

BURN.—ROMAN LITERATURE IN RELATION TO ROMAN ART. By Rev. ROBERT BURN, M.A., late Fellow of Trinity College, Cambridge. Illustrated. Ex. cr. 8vo. 14s.

BURY.—A HISTORY OF THE LATER ROMAN EMPIRE FROM ARCADIUS TO IRENE, A.D. 395-800. By J. B. BURY, M.A., Fellow of Trinity College, Dublin. 2 vols. 8vo. 32s.

BUTCHER.—SOME ASPECTS OF THE GREEK GENIUS. By S. H. BUTCHER, M.A., Professor of Greek, Edinburgh. Cr. 8vo. [*In the Press.*

*****CLASSICAL WRITERS.**—Edited by JOHN RICHARD GREEN, M.A., LL.D. Fcap. 8vo. 1s. 6d. each.
SOPHOCLES. By Prof. L. CAMPBELL, M.A.
EURIPIDES. By Prof. MAHAFFY, D.D.
DEMOSTHENES. By Prof. S. H. BUTCHER, M.A.
VIRGIL. By Prof. NETTLESHIP, M.A.
LIVY. By Rev. W. W. CAPES, M.A.
TACITUS. By Prof. A. J. CHURCH, M.A., and W. J. BRODRIBB, M.A.
MILTON. By Rev. STOPFORD A. BROOKE, M.A.

DYER.—STUDIES OF THE GODS IN GREECE AT CERTAIN SANCTUARIES RECENTLY EXCAVATED. By LOUIS DYER, B.A. Ex. Cr. 8vo. 8s. 6d. net.

ANCIENT HISTORY AND PHILOSOPHY 13

FREEMAN.—Works by EDWARD A. FREEMAN, D.C.L., LL.D., Regius Professor of Modern History in the University of Oxford.
  HISTORY OF ROME. (*Historical Course for Schools.*) 18mo. [*In preparation.*
  HISTORY OF GREECE. (*Historical Course for Schools.*) 18mo. [*In preparation.*
  A SCHOOL HISTORY OF ROME. Cr. 8vo. [*In preparation.*
  HISTORICAL ESSAYS. Second Series. [Greek and Roman History.] 8vo. 10s. 6d.
GARDNER.—SAMOS AND SAMIAN COINS. An Essay. By PERCY GARDNER, Litt.D., Professor of Archæology in the University of Oxford. 8vo. 7s. 6d.
GEDDES.—THE PROBLEM OF THE HOMERIC POEMS. By W. D. GEDDES, Principal of the University of Aberdeen. 8vo. 14s.
GLADSTONE.—Works by the Rt. Hon. W. E. GLADSTONE, M.P.
  THE TIME AND PLACE OF HOMER. Cr. 8vo. 6s. 6d.
  LANDMARKS OF HOMERIC STUDY. Cr. 8vo. 2s. 6d.
  *A PRIMER OF HOMER. 18mo. 1s.
GOW.—A COMPANION TO SCHOOL CLASSICS. By JAMES GOW, Litt.D., Master of the High School, Nottingham. With Illustrations. 2d Ed., revised. Cr. 8vo. 6s.
HARRISON and VERRALL.—MYTHOLOGY AND MONUMENTS OF ANCIENT ATHENS. Translation of a portion of the "Attica" of Pausanias. By MARGARET DE G. VERRALL. With Introductory Essay and Archæological Commentary by JANE E. HARRISON. With Illustrations and Plans. Cr. 8vo. 16s.
JEBB.—Works by R. C. JEBB, Litt.D., Professor of Greek in the University of Cambridge.
  THE ATTIC ORATORS FROM ANTIPHON TO ISAEOS. 2 vols. 8vo. 25s.
  *A PRIMER OF GREEK LITERATURE. 18mo. 1s.
  (See also *Classical Series.*)
KIEPERT.—MANUAL OF ANCIENT GEOGRAPHY. By Dr. H. KIEPERT. Cr. 8vo. 5s.
LANCIANI.—ANCIENT ROME IN THE LIGHT OF RECENT DISCOVERIES. By RODOLFO LANCIANI, Professor of Archæology in the University of Rome. Illustrated. 4to. 24s.
LEAF.—INTRODUCTION TO THE ILIAD FOR ENGLISH READERS. By WALTER LEAF, Litt.D. [*In preparation.*
MAHAFFY.—Works by J. P. MAHAFFY, D.D., Fellow of Trinity College, Dublin, and Professor of Ancient History in the University of Dublin.
  SOCIAL LIFE IN GREECE; from Homer to Menander. Cr. 8vo. 9s.
  GREEK LIFE AND THOUGHT; from the Age of Alexander to the Roman Conquest. Cr. 8vo. 12s. 6d.
  THE GREEK WORLD UNDER ROMAN SWAY. From Plutarch to Polybius. Cr. 8vo. 10s. 6d.
  RAMBLES AND STUDIES IN GREECE. With Illustrations. With Map. Cr. 8vo. 10s. 6d.
  A HISTORY OF CLASSICAL GREEK LITERATURE. Cr. 8vo. Vol. I. In two parts. Part I. Epic and Lyric Poets, with an Appendix on Homer by Prof. SAYCE. Part II. Dramatic Poets. Vol. II. The Prose Writers. In two parts. Part I. Herodotus to Plato. Part II. Isocrates to Aristotle. 4s. 6d. each.
  *A PRIMER OF GREEK ANTIQUITIES. With Illustrations. 18mo. 1s.
  *EURIPIDES. 18mo. 1s. 6d. (*Classical Writers.*)
MAYOR.—BIBLIOGRAPHICAL CLUE TO LATIN LITERATURE. Edited after HÜBNER. By Prof. JOHN E. B. MAYOR. Cr. 8vo. 10s. 6d.
NEWTON.—ESSAYS ON ART AND ARCHÆOLOGY. By Sir CHARLES NEWTON, K.C.B., D.C.L. 8vo. 12s. 6d.
PHILOLOGY.—THE JOURNAL OF PHILOLOGY. Edited by W. A. WRIGHT, M.A., I. BYWATER, M.A., and H. JACKSON, Litt.D. 4s. 6d. each (half-yearly).
SAYCE.—THE ANCIENT EMPIRES OF THE EAST. By A. H. SAYCE, M.A., Deputy-Professor of Comparative Philology, Oxford. Cr. 8vo. 6s.

SCHMIDT and WHITE. AN INTRODUCTION TO THE RHYTHMIC AND METRIC OF THE CLASSICAL LANGUAGES. By Dr. J. H. HEINRICH SCHMIDT. Translated by JOHN WILLIAMS WHITE, Ph.D. 8vo. 10s. 6d.

SHUCHHARDT.—DR. SCHLIEMANN'S EXCAVATIONS AT TROY, TIRYNS, MYCENÆ, ORCHOMENOS, ITHACA, presented in the light of recent knowledge. By Dr. CARL SHUCHHARDT. Translated by EUGENIE SELLERS. Introduction by WALTER LEAF, Litt.D. Illustrated. 8vo. 18s. net.

SHUCKBURGH.—A SCHOOL HISTORY OF ROME. By E. S. SHUCKBURGH, M.A. Cr. 8vo. [In preparation.

*STEWART.—THE TALE OF TROY. Done into English by AUBREY STEWART. Gl. 8vo. 3s. 6d.

*TOZER.—A PRIMER OF CLASSICAL GEOGRAPHY. By H. F. TOZER, M.A. 18mo. 1s.

WALDSTEIN.—CATALOGUE OF CASTS IN THE MUSEUM OF CLASSICAL ARCHÆOLOGY, CAMBRIDGE. By CHARLES WALDSTEIN, University Reader in Classical Archæology. Cr. 8vo. 1s. 6d.
. Also an Edition on Large Paper, small 4to. 5s.

WILKINS.—Works by Prof. WILKINS, Litt.D., LL.D.
*A PRIMER OF ROMAN ANTIQUITIES. Illustrated. 18mo. 1s.
*A PRIMER OF ROMAN LITERATURE. 18mo. 1s.

WILKINS and ARNOLD.—A MANUAL OF ROMAN ANTIQUITIES. By Prof. A. S. WILKINS, Litt.D., and W. T. ARNOLD, M.A. Cr. 8vo. Illustrated. [In preparation.

MODERN LANGUAGES AND LITERATURE.

English; French; German; Modern Greek; Italian; Spanish.

ENGLISH.

*ABBOTT.—A SHAKESPEARIAN GRAMMAR. An Attempt to Illustrate some of the Differences between Elizabethan and Modern English. By the Rev. E. A. ABBOTT, D.D., formerly Headmaster of the City of London School. Ex. fcap. 8vo. 6s.

*BACON.—ESSAYS. With Introduction and Notes, by F. G. SELBY, M.A., Professor of Logic and Moral Philosophy, Deccan College, Poona. Gl. 8vo. 3s.; sewed, 2s. 6d.

*BURKE.—REFLECTIONS ON THE FRENCH REVOLUTION. By the same. Gl. 8vo. 5s.

BROOKE.—*PRIMER OF ENGLISH LITERATURE. By Rev. STOPFORD A. BROOKE, M.A. 18mo. 1s.
EARLY ENGLISH LITERATURE. By the same. 2 vols. 8vo. [Vol. I. In the Press.

BUTLER.—HUDIBRAS. With Introduction and Notes, by ALFRED MILNES, M.A. Ex. fcap. 8vo. Part I. 3s. 6d. Parts II. and III. 4s. 6d.

CAMPBELL.—SELECTIONS. With Introduction and Notes, by CECIL M. BARROW, M.A., Principal of Victoria College, Palghât. Gl. 8vo. [In preparation.

COLLINS.—THE STUDY OF ENGLISH LITERATURE: A Plea for its Recognition and Organisation at the Universities. By J. CHURTON COLLINS, M.A. Cr. 8vo.

COWPER.—*THE TASK: an Epistle to Joseph Hill, Esq.; TIROCINIUM, or a Review of the Schools; and THE HISTORY OF JOHN GILPIN. Edited, with Notes, by W. BENHAM, B.D. Gl. 8vo. 1s. (Globe Readings from Standard Authors.)
THE TASK. With Introduction and Notes, by F. J. ROWE, M.A., and W. T. WEBB, M.A., Professors of English Literature, Presidency College, Calcutta. [In preparation.

*DOWDEN.—A PRIMER OF SHAKESPERE. By Prof. DOWDEN. 18mo. 1s.

DRYDEN.—SELECT PROSE WORKS. Edited, with Introduction and Notes, by Prof. C. D. YONGE. Fcap. 8vo. 2s. 6d.

ENGLISH

*GLOBE READERS. For Standards I.-VI. Edited by A. F. MURISON. Illustrated. Gl. 8vo.

Primer I. (48 pp.)	3d.	Book III. (232 pp.)	1s. 3d.
Primer II. (48 pp.)	3d.	Book IV. (328 pp.)	1s. 9d.
Book I. (132 pp.)	6d.	Book V. (408 pp.)	2s.
Book II. (136 pp.)	9d.	Book VI. (436 pp.)	2s. 6d.

*THE SHORTER GLOBE READERS.—Illustrated. Gl. 8vo.

Primer I. (48 pp.)	3d.	Standard III. (178 pp.)	1s.
Primer II. (48 pp.)	3d.	Standard IV. (182 pp.)	1s.
Standard I. (90 pp.)	6d.	Standard V. (216 pp.)	1s. 3d.
Standard II. (124 pp.)	9d.	Standard VI. (228 pp.)	1s. 6d.

*GOLDSMITH.—THE TRAVELLER, or a Prospect of Society; and THE DESERTED VILLAGE. With Notes, Philological and Explanatory, by J. W. HALES, M.A. Cr. 8vo. 6d.

*THE TRAVELLER AND THE DESERTED VILLAGE. With Introduction and Notes, by A. BARRETT, B.A., Professor of English Literature, Elphinstone College, Bombay. Gl. 8vo. 1s. 9d.; sewed, 1s. 6d. The Traveller (separately), 1s., sewed.

*THE VICAR OF WAKEFIELD. With a Memoir of Goldsmith, by Prof. MASSON. Gl. 8vo. 1s. (Globe Readings from Standard Authors.)

SELECT ESSAYS. With Introduction and Notes, by Prof. C. D. YONGE. Fcap. 8vo. 2s. 6d.

GOSSE.—A HISTORY OF EIGHTEENTH CENTURY LITERATURE (1660-1780). By EDMUND GOSSE, M.A. Cr. 8vo. 7s. 6d.

*GRAY.—POEMS. With Introduction and Notes, by JOHN BRADSHAW, LL.D. Gl. 8vo. 1s. 9d.; sewed, 1s. 6d.

*HALES.—LONGER ENGLISH POEMS. With Notes, Philological and Explanatory, and an Introduction on the Teaching of English, by J. W. HALES, M.A., Professor of English Literature at King's College, London. Ex. fcap. 8vo. 4s. 6d.

*HELPS.—ESSAYS WRITTEN IN THE INTERVALS OF BUSINESS. With Introduction and Notes, by F. J. ROWE, M.A., and W. T. WEBB, M.A. Gl. 8vo. 1s. 9d.; sewed, 1s. 6d.

*JOHNSON.—LIVES OF THE POETS. The Six Chief Lives (Milton, Dryden, Swift, Addison, Pope, Gray), with Macaulay's "Life of Johnson." With Preface and Notes by MATTHEW ARNOLD. Cr. 8vo. 4s. 6d.

KELLNER.—HISTORICAL OUTLINES OF ENGLISH SYNTAX. By L. KELLNER, Ph.D. [In the Press.

*LAMB.—TALES FROM SHAKSPEARE. With Preface by the Rev. CANON AINGER, M.A., LL.D. Gl. 8vo. 2s. (Globe Readings from Standard Authors.)

*LITERATURE PRIMERS.—Edited by JOHN RICHARD GREEN, LL.D. 18mo. 1s. each.

ENGLISH GRAMMAR. By Rev. R. MORRIS, LL.D.
ENGLISH GRAMMAR EXERCISES. By R. MORRIS, LL.D., and H. C. BOWEN, M.A.
EXERCISES ON MORRIS'S PRIMER OF ENGLISH GRAMMAR. By J. WETHERELL, M.A.
ENGLISH COMPOSITION. By Professor NICHOL.
QUESTIONS AND EXERCISES ON ENGLISH COMPOSITION. By Prof. NICHOL and W. S. M'CORMICK.
ENGLISH LITERATURE. By STOPFORD BROOKE, M.A.
SHAKSPERE. By Professor DOWDEN.
THE CHILDREN'S TREASURY OF LYRICAL POETRY. Selected and arranged with Notes by FRANCIS TURNER PALGRAVE. In Two Parts. 1s. each.
PHILOLOGY. By J. PEILE, Litt.D.
ROMAN LITERATURE. By Prof. A. S. WILKINS, Litt.D.
GREEK LITERATURE. By Prof. JEBB, Litt.D.
HOMER. By the Rt. Hon. W. E. GLADSTONE, M.P.

A HISTORY OF ENGLISH LITERATURE IN FOUR VOLUMES. Cr. 8vo.
EARLY ENGLISH LITERATURE. By STOPFORD BROOKE, M.A. [In preparation.

MODERN LANGUAGES AND LITERATURE

ELIZABETHAN LITERATURE. (1560-1665.) By GEORGE SAINTSBURY. 7s. 6d.
EIGHTEENTH CENTURY LITERATURE. (1660-1780.) By EDMUND GOSSE, M.A. 7s. 6d.
THE MODERN PERIOD. By Prof. DOWDEN. [*In preparation.*]

*MACMILLAN'S READING BOOKS.

PRIMER. 18mo. 48 pp. 2d.
BOOK I. for Standard I. 96 pp. 4d.
BOOK II. for Standard II. 144 pp. 5d.
BOOK III. for Standard III. 160 pp. 6d.
BOOK IV. for Standard IV. 176 pp. 8d.
BOOK V. for Standard V. 380 pp. 1s.
BOOK VI. for Standard VI. Cr. 8vo. 430 pp. 2s.

Book VI. is fitted for Higher Classes, and as an Introduction to English Literature.

*MACMILLAN'S COPY BOOKS.—1. Large Post 4to. Price 4d. each. 2. Post Oblong. Price 2d. each.
1. INITIATORY EXERCISES AND SHORT LETTERS.
2. WORDS CONSISTING OF SHORT LETTERS.
3. LONG LETTERS. With Words containing Long Letters—Figures.
4. WORDS CONTAINING LONG LETTERS.
4a. PRACTISING AND REVISING COPY-BOOK. For Nos. 1 to 4.
5. CAPITALS AND SHORT HALF-TEXT. Words beginning with a Capital.
6. HALF-TEXT WORDS beginning with Capitals—Figures.
7. SMALL-HAND AND HALF-TEXT. With Capitals and Figures.
8. SMALL-HAND AND HALF-TEXT. With Capitals and Figures.
8a. PRACTISING AND REVISING COPY-BOOK. For Nos. 5 to 8.
9. SMALL-HAND SINGLE HEADLINES—Figures.
10. SMALL-HAND SINGLE HEADLINES—Figures.
11. SMALL-HAND DOUBLE HEADLINES—Figures.
12. COMMERCIAL AND ARITHMETICAL EXAMPLES, &c.
12a. PRACTISING AND REVISING COPY-BOOK. For Nos. 8 to 12.

Nos. 3, 4, 5, 6, 7, 8, 9 *may be had with Goodman's Patent Sliding Copies.* Large Post 4to. Price 6d. each.

MARTIN.—*THE POET'S HOUR: Poetry selected and arranged for Children. By FRANCES MARTIN. 18mo. 2s. 6d.
*SPRING-TIME WITH THE POETS. By the same. 18mo. 3s. 6d.

*MILTON.—PARADISE LOST. Books I. and II. With Introduction and Notes, by MICHAEL MACMILLAN, B.A., Professor of Logic and Moral Philosophy, Elphinstone College, Bombay. Gl. 8vo. 1s. 9d.; sewed, 1s. 6d. Or separately, 1s. 3d.; sewed, 1s. each.
*L'ALLEGRO, IL PENSEROSO, LYCIDAS, ARCADES, SONNETS, &c. With Introduction and Notes, by W. BELL, M.A., Professor of Philosophy and Logic, Government College, Lahore. Gl. 8vo. 1s. 9d.; sewed, 1s. 6d.
*COMUS. By the same. Gl. 8vo. 1s. 3d.; sewed, 1s.
*SAMSON AGONISTES. By H. M. PERCIVAL, M.A., Professor of English Literature, Presidency College, Calcutta. Gl. 8vo. 2s.; sewed, 1s. 9d.
*INTRODUCTION TO THE STUDY OF MILTON. By STOPFORD BROOKE M.A. Fcap. 8vo. 1s. 6d. (*Classical Writers.*)

MORRIS.—Works by the Rev. R. MORRIS, LL.D.
*PRIMER OF ENGLISH GRAMMAR. 18mo. 1s.
*ELEMENTARY LESSONS IN HISTORICAL ENGLISH GRAMMAR, containing Accidence and Word-Formation. 18mo. 2s. 6d.
*HISTORICAL OUTLINES OF ENGLISH ACCIDENCE, comprising Chapters on the History and Development of the Language, and on Word-Formation. Ex. fcap. 8vo. 6s.

NICHOL and M'CORMICK.—A SHORT HISTORY OF ENGLISH LITERATURE. By Prof. JOHN NICHOL and Prof. W. S. M'CORMICK. [*In preparation.*]

OLIPHANT.—THE OLD AND MIDDLE ENGLISH. By T. L. KINGTON OLIPHANT. New Ed., revised and enlarged, of "The Sources of Standard English." 2nd Ed. Gl. 8vo. 9s.
THE NEW ENGLISH. By the same. 2 vols. Cr. 8vo. 21s.

ENGLISH 17

PALGRAVE.—THE CHILDREN'S TREASURY OF LYRICAL POETRY. Selected and arranged, with Notes, by FRANCIS T. PALGRAVE. 18mo. 2s. 6d. Also in Two Parts. 1s. each.

PATMORE.—THE CHILDREN'S GARLAND FROM THE BEST POETS. Selected and arranged by COVENTRY PATMORE. Gl. 8vo. 2s. (*Globe Readings from Standard Authors.*)

PLUTARCH.—Being a Selection from the Lives which illustrate Shakespeare. North's Translation. Edited, with Introductions, Notes, Index of Names, and Glossarial Index, by Prof. W. W. SKEAT, Litt.D. Cr. 8vo. 6s.

RANSOME.—SHORT STUDIES OF SHAKESPEARE'S PLOTS. By CYRIL RANSOME, Professor of Modern History and Literature, Yorkshire College, Leeds. Cr. 8vo. 3s. 6d.

RYLAND.—CHRONOLOGICAL OUTLINES OF ENGLISH LITERATURE. By F. RYLAND, M.A. Cr. 8vo. 6s.

SAINTSBURY.—A HISTORY OF ELIZABETHAN LITERATURE. 1500-1665. By GEORGE SAINTSBURY. Cr. 8vo. 7s. 6d.

SCOTT.—*LAY OF THE LAST MINSTREL, and THE LADY OF THE LAKE. Edited, with Introduction and Notes, by FRANCIS TURNER PALGRAVE. Gl. 8vo. 1s. (*Globe Readings from Standard Authors.*)

**THE LAY OF THE LAST MINSTREL. With Introduction and Notes, by G. H. STUART, M.A., and E. H. ELLIOT, B.A. Gl. 8vo. 2s.; sewed, 1s. 9d. Introduction and Canto I. 9d. sewed. Cantos I. to III. 1s. 3d.; sewed, 1s. Cantos IV. to VI. 1s. 3d.; sewed, 1s.

**MARMION, and THE LORD OF THE ISLES. By F. T. PALGRAVE. Gl. 8vo. 1s. (*Globe Readings from Standard Authors.*)

**MARMION. With Introduction and Notes, by MICHAEL MACMILLAN, B.A. Gl. 8vo. 3s.; sewed, 2s. 6d.

**THE LADY OF THE LAKE. By G. H. STUART, M.A. Gl. 8vo. 2s. 6d.; sewed, 2s.

**ROKEBY. With Introduction and Notes, by MICHAEL MACMILLAN, B.A. Gl. 8vo. 3s.; sewed, 2s. 6d.

SHAKESPEARE.—*A SHAKESPEARIAN GRAMMAR. By Rev. E. A. ABBOTT, D.D. Gl. 8vo. 6s.

A SHAKESPEARE MANUAL. By F. G. FLEAY, M.A. 2d Ed. Ex. fcap. 8vo. 4s. 6d.

*A PRIMER OF SHAKESPERE. By Prof. DOWDEN. 18mo. 1s.

*SHORT STUDIES OF SHAKESPEARE'S PLOTS. By CYRIL RANSOME, M.A. Cr. 8vo. 3s. 6d.

**THE TEMPEST. With Introduction and Notes, by K. DEIGHTON, late Principal of Agra College. Gl. 8vo. 1s. 9d.; sewed, 1s. 6d.

**MUCH ADO ABOUT NOTHING. By the same. Gl. 8vo. 2s.; sewed, 1s. 9d.

**A MIDSUMMER NIGHT'S DREAM. By the same. Gl. 8vo. 1s. 9d.; sewed, 1s. 6d.

**THE MERCHANT OF VENICE. By the same. Gl. 8vo. 1s. 9d.; sewed, 1s. 6d.

*AS YOU LIKE IT. By the same. Gl. 8vo. 1s. 9d.; sewed, 1s. 6d.

*TWELFTH NIGHT. By the same. Gl. 8vo. 1s. 9d.; sewed, 1s. 6d.

**THE WINTER'S TALE. By the same. Gl. 8vo. 2s.; sewed, 1s. 9d.

*KING JOHN. By the same. Gl. 8vo. 1s. 9d.; sewed, 1s. 6d.

*RICHARD II. By the same. Gl. 8vo. 1s. 9d.; sewed, 1s. 6d.

*HENRY V. By the same. Gl. 8vo. 1s. 9d.; sewed, 1s. 6d.

*RICHARD III. By C. H. TAWNEY, M.A., Principal and Professor of English Literature, Presidency College, Calcutta. Gl. 8vo. 2s. 6d.; sewed, 2s.

*CORIOLANUS. By K. DEIGHTON. Gl. 8vo. 2s. 6d.; sewed, 2s.

*JULIUS CÆSAR. By the same. Gl. 8vo. 1s. 9d.; sewed, 1s. 6d.

*MACBETH. By the same. Gl. 8vo. 1s. 9d.; sewed, 1s. 6d.

*HAMLET. By the same. Gl. 8vo. 2s. 6d.; sewed, 2s.

*KING LEAR. By the same. Gl. 8vo. 1s. 9d.; sewed, 1s. 6d.

*OTHELLO. By the same. Gl. 8vo. 2s.; sewed, 1s. 9d.

*ANTONY AND CLEOPATRA. By the same. Gl. 8vo. 2s. 6d.; sewed, 2s.
*CYMBELINE. By the same. Gl. 8vo. 2s. 6d.; sewed, 2s.
*SONNENSCHEIN and MEIKLEJOHN.—THE ENGLISH METHOD OF TEACHING TO READ. By A. SONNENSCHEIN and J. M. D. MEIKLEJOHN, M.A. Fcap. 8vo.
THE NURSERY BOOK, containing all the Two-Letter Words in the Language. 1d. (Also in Large Type on Sheets for School Walls. 5s.)
THE FIRST COURSE, consisting of Short Vowels with Single Consonants. 7d.
THE SECOND COURSE, with Combinations and Bridges, consisting of Short Vowels with Double Consonants. 7d.
THE THIRD AND FOURTH COURSES, consisting of Long Vowels, and all the Double Vowels in the Language. 7d.
*SOUTHEY.—LIFE OF NELSON. With Introduction and Notes, by MICHAEL MACMILLAN, B.A. Gl. 8vo. 3s.; sewed, 2s. 6d.
SPENSER.—FAIRY QUEEN. Book I. With Introduction and Notes, by H. M. PERCIVAL, M.A. [*In the Press.*
TAYLOR.—WORDS AND PLACES; or, Etymological Illustrations of History, Ethnology, and Geography. By Rev. ISAAC TAYLOR, Litt.D. With Maps. Gl. 8vo. 6s.
TENNYSON.—THE COLLECTED WORKS OF LORD TENNYSON. An Edition for Schools. In Four Parts. Cr. 8vo. 2s. 6d. each.
TENNYSON FOR THE YOUNG. Edited, with Notes for the Use of Schools, by the Rev. ALFRED AINGER, LL.D., Canon of Bristol. 18mo. 1s. net.
[*In the Press.*
*SELECTIONS FROM TENNYSON. With Introduction and Notes, by F. J. ROWE, M.A., and W. T. WEBB, M.A. Gl. 8vo. 3s. 6d.
This selection contains:—Recollections of the Arabian Nights, The Lady of Shalott, Œnone, The Lotos Eaters, Ulysses, Tithonus, Morte d'Arthur, Sir Galahad, Dora, Ode on the Death of the Duke of Wellington, and The Revenge.
*ENOCH ARDEN. By W. T. WEBB, M.A. Gl. 8vo. 2s.
*AYLMER'S FIELD. By W. T. WEBB, M.A. 2s.
THE PRINCESS; A MEDLEY. By P. M. WALLACE, B.A. [*In the Press.*
*THE COMING OF ARTHUR, AND THE PASSING OF ARTHUR. By F. J. ROWE, M.A. Gl. 8vo. 2s.
THRING.—THE ELEMENTS OF GRAMMAR TAUGHT IN ENGLISH. By EDWARD THRING, M.A. With Questions. 4th Ed. 18mo. 2s.
*VAUGHAN.—WORDS FROM THE POETS. By C. M. VAUGHAN. 18mo. 1s.
WARD.—THE ENGLISH POETS. Selections, with Critical Introductions by various Writers and a General Introduction by MATTHEW ARNOLD. Edited by T. H. WARD, M.A. 4 Vols. Vol. I. CHAUCER TO DONNE.—Vol. II. BEN JONSON TO DRYDEN.—Vol. III. ADDISON TO BLAKE.—Vol. IV. WORDSWORTH TO ROSSETTI. 2d Ed. Cr. 8vo. 7s. 6d. each.
*WETHERELL.—EXERCISES ON MORRIS'S PRIMER OF ENGLISH GRAMMAR. By JOHN WETHERELL, M.A., Headmaster of Towcester Grammar School. 18mo. 1s.
WOODS.—*A FIRST POETRY BOOK. By M. A. WOODS, Head Mistress of the Clifton High School for Girls. Fcap. 8vo. 2s. 6d.
*A SECOND POETRY BOOK. By the same. In Two Parts. 2s. 6d. each.
*A THIRD POETRY BOOK. By the same. 4s. 6d.
HYMNS FOR SCHOOL WORSHIP. By the same. 18mo. 1s. 6d.
WORDSWORTH.—SELECTIONS. With Introduction and Notes, by F. J. ROWE, M.A., and W. T. WEBB, M.A. Gl. 8vo. [*In preparation.*
YONGE.—*A BOOK OF GOLDEN DEEDS. By CHARLOTTE M. YONGE. Gl. 8vo. 2s.
*THE ABRIDGED BOOK OF GOLDEN DEEDS. 18mo. 1s.

FRENCH.

BEAUMARCHAIS.—LE BARBIER DE SEVILLE. With Introduction and Notes. By L. P. BLOUET. Fcap. 8vo. 3s. 6d.
*BOWEN.—FIRST LESSONS IN FRENCH. By H. COURTHOPE BOWEN, M.A. Ex. fcap. 8vo. 1s.

FRENCH

BREYMANN.—Works by HERMANN BREYMANN, Ph.D., Professor of Philology in the University of Munich.
  FIRST FRENCH EXERCISE BOOK. Ex. fcap. 8vo. 4s. 6d.
  SECOND FRENCH EXERCISE BOOK. Ex. fcap. 8vo. 2s. 6d.
FASNACHT.—Works by G. E. FASNACHT, late Assistant Master at Westminster.
  THE ORGANIC METHOD OF STUDYING LANGUAGES. Ex. fcap. 8vo. I. French. 3s. 6d.
  A SYNTHETIC FRENCH GRAMMAR FOR SCHOOLS. Cr. 8vo. 3s. 6d.
  GRAMMAR AND GLOSSARY OF THE FRENCH LANGUAGE OF THE SEVENTEENTH CENTURY. Cr. 8vo. [In preparation.
MACMILLAN'S PRIMARY SERIES OF FRENCH READING BOOKS.—Edited by G. E. FASNACHT. With Illustrations, Notes, Vocabularies, and Exercises. Gl. 8vo.
  *FRENCH READINGS FOR CHILDREN. By G. E. FASNACHT. 1s. 6d.
  *CORNAZ—NOS ENFANTS ET LEURS AMIS. By EDITH HARVEY. 1s. 6d.
  *DE MAISTRE—LA JEUNE SIBÉRIENNE ET LE LÉPREUX DE LA CITÉ D'AOSTE. By STEPHANE BARLET, B.Sc. etc. 1s. 6d.
  *FLORIAN—FABLES. By Rev. CHARLES YELD, M.A., Headmaster of University School, Nottingham. 1s. 6d.
  *LA FONTAINE—A SELECTION OF FABLES. By L. M. MORIARTY, B.A., Assistant Master at Harrow. 2s. 6d.
  *MOLESWORTH—FRENCH LIFE IN LETTERS. By Mrs. MOLESWORTH. 1s. 6d.
  *PERRAULT—CONTES DE FÉES. By G. E. FASNACHT. 1s. 6d.
MACMILLAN'S PROGRESSIVE FRENCH COURSE.—By G. E. FASNACHT. Ex. fcap. 8vo.
  *FIRST YEAR, containing Easy Lessons on the Regular Accidence. 1s.
  *SECOND YEAR, containing an Elementary Grammar with copious Exercises, Notes, and Vocabularies. 2s.
  *THIRD YEAR, containing a Systematic Syntax, and Lessons in Composition. 2s. 6d.
  THE TEACHER'S COMPANION TO MACMILLAN'S PROGRESSIVE FRENCH COURSE. With Copious Notes, Hints for Different Renderings, Synonyms, Philological Remarks, etc. By G. E. FASNACHT. Ex. fcap. 8vo. Each Year 4s. 6d.
*MACMILLAN'S FRENCH COMPOSITION.—By G. E. FASNACHT. Ex. fcap. 8vo. Part I. Elementary. 2s. 6d. Part II. Advanced. [In the Press.
  THE TEACHER'S COMPANION TO MACMILLAN'S COURSE OF FRENCH COMPOSITION. By G. E. FASNACHT. Part I. Ex. fcap. 8vo. 4s. 6d.
MACMILLAN'S PROGRESSIVE FRENCH READERS. By G. E. FASNACHT. Ex. fcap. 8vo.
  *FIRST YEAR, containing Tales, Historical Extracts, Letters, Dialogues, Ballads, Nursery Songs, etc., with Two Vocabularies: (1) in the order of subjects; (2) in alphabetical order. With Imitative Exercises. 2s. 6d.
  *SECOND YEAR, containing Fiction in Prose and Verse, Historical and Descriptive Extracts, Essays, Letters, Dialogues, etc. With Imitative Exercises. 2s. 6d.
MACMILLAN'S FOREIGN SCHOOL CLASSICS. Edited by G. E. FASNACHT. 18mo.
  *CORNEILLE—LE CID. By G. E. FASNACHT. 1s.
  *DUMAS—LES DEMOISELLES DE ST. CYR. By VICTOR OGER, Lecturer at University College, Liverpool. 1s. 6d.
  LA FONTAINE'S FABLES. Books I.-VI. By L. M. MORIARTY, B.A., Assistant Master at Harrow. [In preparation.
  *MOLIÈRE—L'AVARE. By the same. 1s.
  *MOLIÈRE—LE BOURGEOIS GENTILHOMME. By the same. 1s. 6d.
  *MOLIÈRE—LES FEMMES SAVANTES. By G. E. FASNACHT. 1s.
  *MOLIÈRE—LE MISANTHROPE. By the same. 1s.

*MOLIÈRE—LE MÉDECIN MALGRÉ LUI. By the same. 1s.
*MOLIÈRE—LES PRÉCIEUSES RIDICULES. By the same. 1s.
*RACINE—BRITANNICUS. By E. PELLISSIER, M.A. 2s.
*FRENCH READINGS FROM ROMAN HISTORY. Selected from various Authors, by C. COLBECK, M.A., Assistant Master at Harrow. 4s. 6d.
*SAND, GEORGE—LA MARE AU DIABLE. By W. E. RUSSELL, M.A., Assistant Master at Haileybury. 1s.
*SANDEAU, JULES—MADEMOISELLE DE LA SEIGLIÈRE. By H. C. STEEL, Assistant Master at Winchester. 1s. 6d.
*VOLTAIRE—CHARLES XII. By G. E. FASNACHT. 3s. 6d.
*MASSON.—A COMPENDIOUS DICTIONARY OF THE FRENCH LANGUAGE. Adapted from the Dictionaries of Professor A. ELWALL. By GUSTAVE MASSON. Cr. 8vo. 3s. 6d.
MOLIÈRE.—LE MALADE IMAGINAIRE. With Introduction and Notes, by F. TARVER, M.A., Assistant Master at Eton. Fcap. 8vo. 2s. 6d.
*PELLISSIER.—FRENCH ROOTS AND THEIR FAMILIES. A Synthetic Vocabulary, based upon Derivations. By E. PELLISSIER, M.A., Assistant Master at Clifton College. Gl. 8vo. 6s.

GERMAN.

BEHAGHEL.—THE GERMAN LANGUAGE. By Dr. OTTO BEHAGHEL. Translated by EMIL TRECHMANN, B.A., Ph.D., Lecturer in Modern Literature in the University of Sydney, N.S.W. Gl. 8vo. 3s. 6d.
HUSS.—A SYSTEM OF ORAL INSTRUCTION IN GERMAN, by means of Progressive Illustrations and Applications of the leading Rules of Grammar. By H. C. O. HUSS, Ph.D. Cr. 8vo. 5s.
MACMILLAN'S PRIMARY SERIES OF GERMAN READING BOOKS. Edited by G. E. FASNACHT. With Notes, Vocabularies, and Exercises. Gl. 8vo.
*GRIMM—KINDER UND HAUSMÄRCHEN. By G. E. FASNACHT. 2s. 6d.
*HAUFF—DIE KARAVANE. By HERMAN HAGER, Ph.D., Lecturer in the Owens College, Manchester. 3s.
*SCHMID, CHR. VON—H. VON EICHENFELS. By G. E. FASNACHT. 2s. 6d.
MACMILLAN'S PROGRESSIVE GERMAN COURSE. By G. E. FASNACHT. Ex. fcap. 8vo.
　*FIRST YEAR. Easy lessons and Rules on the Regular Accidence. 1s. 6d.
　*SECOND YEAR. Conversational Lessons in Systematic Accidence and Elementary Syntax. With Philological Illustrations and Etymological Vocabulary. 3s. 6d.
　THIRD YEAR. [In the Press.
TEACHER'S COMPANION TO MACMILLAN'S PROGRESSIVE GERMAN COURSE. With copious Notes, Hints for Different Renderings, Synonyms, Philological Remarks, etc. By G. E. FASNACHT. Ex. fcap. 8vo. FIRST YEAR. 4s. 6d. SECOND YEAR. 4s. 6d.
MACMILLAN'S GERMAN COMPOSITION. By G. E. FASNACHT. Ex. fcap. 8vo.
　*I. FIRST COURSE Parallel German-English Extracts and Parallel English-German Syntax. 2s. 6d.
TEACHER'S COMPANION TO MACMILLAN'S GERMAN COMPOSITION. By G. E. FASNACHT. FIRST COURSE. Gl. 8vo. 4s. 6d.
MACMILLAN'S PROGRESSIVE GERMAN READERS. By G. E. FASNACHT. Ex. fcap. 8vo.
　*FIRST YEAR, containing an Introduction to the German order of Words, with Copious Examples, extracts from German Authors in Prose and Poetry; Notes, and Vocabularies. 2s. 6d.
MACMILLAN'S FOREIGN SCHOOL CLASSICS.—Edited by G. E. FASNACHT. 18mo.
　FREYTAG (G.)—DOKTOR LUTHER. By F. STORR, M.A., Headmaster of the Modern Side, Merchant Taylors' School. [In preparation.

GERMAN—MODERN GREEK—ITALIAN—SPANISH

*GOETHE—GÖTZ VON BERLICHINGEN. By H. A. BULL, M.A., Assistant Master at Wellington. 2s.
*GOETHE—FAUST. PART I., followed by an Appendix on PART II. By JANE LEE, Lecturer in German Literature at Newnham College, Cambridge. 4s. 6d.
*HEINE—SELECTIONS FROM THE REISEBILDER AND OTHER PROSE WORKS. By C. COLBECK, M.A., Assistant Master at Harrow. 2s. 6d.
LESSING—MINNA VON BARNHELM. By JAMES SIME, M.A. [In preparation.
*SCHILLER—SELECTIONS FROM SCHILLER'S LYRICAL POEMS. With a Memoir of Schiller. By E. J. TURNER, B.A., and E. D. A. MORSHEAD, M.A., Assistant Masters at Winchester. 2s. 6d.
*SCHILLER—DIE JUNGFRAU VON ORLEANS. By JOSEPH GOSTWICK. 2s. 6d.
*SCHILLER—MARIA STUART. By C. SHELDON, D.Litt., of the Royal Academical Institution, Belfast. 2s. 6d.
*SCHILLER—WILHELM TELL. By G. E. FASNACHT. 2s. 6d.
*SCHILLER—WALLENSTEIN. Part I. DAS LAGER. By H. B. COTTERILL, M.A. 2s.
*UHLAND—SELECT BALLADS. Adapted as a First Easy Reading Book for Beginners. With Vocabulary. By G. E. FASNACHT. 1s.
*PYLODET.—NEW GUIDE TO GERMAN CONVERSATION; containing an Alphabetical List of nearly 800 Familiar Words; followed by Exercises, Vocabulary of Words in frequent use, Familiar Phrases and Dialogues, a Sketch of German Literature, Idiomatic Expressions, etc. By L. PYLODET. 18mo. 2s. 6d.
SMITH.—COMMERCIAL GERMAN. By F. C. SMITH, M.A. [In the Press.
WHITNEY.—A COMPENDIOUS GERMAN GRAMMAR. By W. D. WHITNEY, Professor of Sanskrit and Instructor in Modern Languages in Yale College. Cr. 8vo. 4s. 6d.
A GERMAN READER IN PROSE AND VERSE. By the same. With Notes and Vocabulary. Cr. 8vo. 5s.
*WHITNEY and EDGREN.—A COMPENDIOUS GERMAN AND ENGLISH DICTIONARY, with Notation of Correspondences and Brief Etymologies. By Prof. W. D. WHITNEY, assisted by A. H. EDGREN. Cr. 8vo. 7s. 6d.
THE GERMAN-ENGLISH PART, separately, 5s.

MODERN GREEK.

VINCENT and DICKSON.—HANDBOOK TO MODERN GREEK. By Sir EDGAR VINCENT, K.C.M.G., and T. G. DICKSON, M.A. With Appendix on the relation of Modern and Classical Greek by Prof. JEBB. Cr. 8vo. 6s.

ITALIAN.

DANTE.—THE INFERNO OF DANTE. With Translation and Notes, by A. J. BUTLER, M.A. Cr. 8vo. [In the Press.
THE PURGATORIO OF DANTE. With Translations and Notes, by the same. Cr. 8vo. 12s. 6d.
THE PARADISO OF DANTE. With Translation and Notes, by the same. 2d. Ed. Cr. 8vo. 12s. 6d.
READINGS ON THE PURGATORIO OF DANTE. Chiefly based on the Commentary of Benvenuto Da Imola. By the Hon. W. WARREN VERNON, M.A. With an Introduction by the Very Rev. the DEAN OF ST. PAUL'S. 2 vols. Cr. 8vo. 24s.

SPANISH.

CALDERON.—FOUR PLAYS OF CALDERON. With Introduction and Notes. By NORMAN MACCOLL, M.A. Cr. 8vo. 14s.
The four plays here given are *El Principe Constante, La Vida es Sueño, El Alcalde de Zalamea,* and *El Escondido y La Tapada.*

MATHEMATICS.

Arithmetic, Book-keeping, Algebra, Euclid and Pure Geometry, Geometrical Drawing, Mensuration, Trigonometry, Analytical Geometry (Plane and Solid), Problems and Questions in Mathematics, Higher Pure Mathematics, Mechanics (Statics, Dynamics, Hydrostatics, Hydrodynamics: see also Physics), Physics (Sound, Light, Heat, Electricity, Elasticity, Attractions, &c.), Astronomy, Historical.

ARITHMETIC.

*ALDIS.—THE GREAT GIANT ARITHMOS. A most Elementary Arithmetic for Children. By MARY STEADMAN ALDIS. Illustrated. Gl. 8vo. 2s. 6d.

ARMY PRELIMINARY EXAMINATION, SPECIMENS OF PAPERS SET AT THE, 1882-90.—With Answers to the Mathematical Questions. Subjects: Arithmetic, Algebra, Euclid, Geometrical Drawing, Geography, French, English Dictation. Cr. 8vo. 3s. 6d.

*BRADSHAW.—A COURSE OF EASY ARITHMETICAL EXAMPLES FOR BEGINNERS. By J. G. BRADSHAW, B.A., Assistant Master at Clifton College. Gl. 8vo. 2s. With Answers, 2s. 6d.

*BROOKSMITH.—ARITHMETIC IN THEORY AND PRACTICE. By J. BROOKSMITH, M.A. Cr. 8vo. 4s. 6d. KEY. Crown 8vo. 10s. 6d.

*BROOKSMITH.—ARITHMETIC FOR BEGINNERS. By J. and E. J. BROOKSMITH. Gl. 8vo. 1s. 6d.

CANDLER.—HELP TO ARITHMETIC. Designed for the use of Schools. By H. CANDLER, Mathematical Master of Uppingham School. 2d Ed. Ex. fcap. 8vo. 2s. 6d.

*DALTON.—RULES AND EXAMPLES IN ARITHMETIC. By the Rev. T. DALTON, M.A., Senior Mathematical Master at Eton. New Ed., with Answers. 18mo. 2s. 6d.

*GOYEN.—HIGHER ARITHMETIC AND ELEMENTARY MENSURATION. By P. GOYEN, Inspector of Schools, Dunedin, New Zealand. Cr. 8vo. 5s.

*HALL and KNIGHT.—ARITHMETICAL EXERCISES AND EXAMINATION PAPERS. With an Appendix containing Questions in LOGARITHMS and MENSURATION. By H. S. HALL, M.A., Master of the Military and Engineering Side, Clifton College, and S. R. KNIGHT, B.A. Gl. 8vo. 2s. 6d.

LOCK.—Works by Rev. J. B. LOCK, M.A., Senior Fellow and Bursar of Gonville and Caius College, Cambridge.

*ARITHMETIC FOR SCHOOLS. With Answers and 1000 additional Examples for Exercise. 8d Ed., revised. Gl. 8vo. 4s. 6d. Or, Part I. 2s. Part II. 3s. KEY. Cr. 8vo. 10s. 6d.

*ARITHMETIC FOR BEGINNERS. A School Class-Book of Commercial Arithmetic. Gl. 8vo. 2s. 6d. KEY. Cr. 8vo. 8s. 6d.

*A SHILLING BOOK OF ARITHMETIC, FOR ELEMENTARY SCHOOLS. 18mo. 1s. With Answers. 1s. 6d.

*PEDLEY.—EXERCISES IN ARITHMETIC for the Use of Schools. Containing more than 7000 original Examples. By SAMUEL PEDLEY. Cr. 8vo. 5s. Also in Two Parts, 2s. 6d. each.

SMITH.—Works by Rev. BARNARD SMITH, M.A., late Fellow and Senior Bursar of St. Peter's College, Cambridge.

ARITHMETIC AND ALGEBRA, in their Principles and Application; with numerous systematically arranged Examples taken from the Cambridge Examination Papers, with especial reference to the Ordinary Examination for the B.A. Degree. New Ed., carefully revised. Cr. 8vo. 10s. 6d.

*ARITHMETIC FOR SCHOOLS. Cr. 8vo. 4s. 6d. KEY. Cr. 8vo. 8s. 6d. New Edition. Revised by Prof. W. H. HUDSON. [*In preparation.*

BOOK-KEEPING—ALGEBRA 23

EXERCISES IN ARITHMETIC. Cr. 8vo. 2s. With Answers, 2s. 6d. Answers separately, 6d.

SCHOOL CLASS-BOOK OF ARITHMETIC. 18mo. 3s. Or separately, in Three Parts, 1s. each. KEYS. Parts I., II., and III., 2s. 6d. each.

SHILLING BOOK OF ARITHMETIC. 18mo. Or separately, Part I., 2d.; Part II., 3d.; Part III., 7d. Answers, 6d. KEY. 18mo. 4s. 6d.

*THE SAME, with Answers. 18mo, cloth. 1s. 6d.

EXAMINATION PAPERS IN ARITHMETIC. 18mo. 1s. 6d. The Same, with Answers. 18mo. 2s. Answers, 6d. KEY. 18mo. 4s. 6d.

THE METRIC SYSTEM OF ARITHMETIC, ITS PRINCIPLES AND APPLICATIONS, with Numerous Examples. 18mo. 3d.

A CHART OF THE METRIC SYSTEM, on a Sheet, size 42 in. by 34 in. on Roller. 3s. 6d. Also a Small Chart on a Card. Price 1d.

EASY LESSONS IN ARITHMETIC, combining Exercises in Reading, Writing, Spelling, and Dictation. Part I. Cr. 8vo. 9d.

EXAMINATION CARDS IN ARITHMETIC. With Answers and Hints. Standards I. and II., in box, 1s. Standards III., IV., and V., in boxes, 1s. each. Standard VI. in Two Parts, in boxes, 1s. each.

A and B papers, of nearly the same difficulty, are given so as to prevent copying, and the colours of the A and B papers differ in each Standard, and from those of every other Standard, so that a master or mistress can see at a glance whether the children have the proper papers.

BOOK-KEEPING.

*THORNTON.—FIRST LESSONS IN BOOK-KEEPING. By J. THORNTON. Cr. 8vo. 2s. 6d. KEY. Oblong 4to. 10s. 6d.

*PRIMER OF BOOK-KEEPING. 18mo. 1s. KEY. Demy 8vo. 2s. 6d.

ALGEBRA.

*DALTON.—RULES AND EXAMPLES IN ALGEBRA. By Rev. T. DALTON, Senior Mathematical Master at Eton. Part I. 18mo. 2s. KEY. Cr. 8vo. 7s. 6d. Part II. 18mo. 2s. 6d.

HALL and KNIGHT.—Works by H. S. HALL, M.A., Master of the Military and Engineering Side, Clifton College, and S. R. KNIGHT, B.A.

*ELEMENTARY ALGEBRA FOR SCHOOLS. 6th Ed., revised and corrected. Gl. 8vo, bound in maroon coloured cloth, 3s. 6d.; with Answers, bound in green coloured cloth, 4s. 6d. KEY. 8s. 6d.

*ALGEBRAICAL EXERCISES AND EXAMINATION PAPERS. To accompany ELEMENTARY ALGEBRA. 2d Ed., revised. Gl. 8vo. 2s. 6d.

*HIGHER ALGEBRA. 4th Ed. Cr. 8vo. 7s. 6d. KEY. Cr. 8vo. 10s. 6d.

*JONES and CHEYNE.—ALGEBRAICAL EXERCISES. Progressively Arranged. By Rev. C. A. JONES and C. H. CHEYNE, M.A., late Mathematical Masters at Westminster School. 18mo. 2s. 6d.

KEY. By Rev. W. FAILES, M.A., Mathematical Master at Westminster School. Cr. 8vo. 7s. 6d.

SMITH.—ARITHMETIC AND ALGEBRA, in their Principles and Application; with numerous systematically arranged Examples taken from the Cambridge Examination Papers, with especial reference to the Ordinary Examination for the B.A. Degree. By Rev. BARNARD SMITH, M.A. New Edition, carefully revised. Cr. 8vo. 10s. 6d.

SMITH.—Works by CHARLES SMITH, M.A., Master of Sidney Sussex College, Cambridge.

*ELEMENTARY ALGEBRA. 2d Ed., revised. Gl. 8vo. 4s. 6d. KEY. By A. G. CRACKNELL, B.A. Cr. 8vo. 10s. 6d.

*A TREATISE ON ALGEBRA. 2d Ed. Cr. 8vo. 7s. 6d. KEY. Cr. 8vo. 10s. 6d.

TODHUNTER.—Works by ISAAC TODHUNTER, F.R.S.

*ALGEBRA FOR BEGINNERS. 18mo. 2s. 6d. KEY. Cr. 8vo. 6s. 6d.

24   MATHEMATICS

*ALGEBRA FOR COLLEGES AND SCHOOLS. By Isaac Todhunter, F.R.S. Cr. 8vo. 7s. 6d. KEY. Cr. 8vo. 10s. 6d.

EUCLID AND PURE GEOMETRY.

COCKSHOTT and WALTERS.—A TREATISE ON GEOMETRICAL CONICS. In accordance with the Syllabus of the Association for the Improvement of Geometrical Teaching. By A. Cockshott, M.A., Assistant Master at Eton, and Rev. F. B. Walters, M.A., Principal of King William's College, Isle of Man. Cr. 8vo. 5s.

CONSTABLE.—GEOMETRICAL EXERCISES FOR BEGINNERS. By Samuel Constable. Cr. 8vo. 3s. 6d.

CUTHBERTSON.—EUCLIDIAN GEOMETRY. By Francis Cuthbertson, M.A., LL.D. Ex. fcap. 8vo. 4s. 6d.

DAY.—PROPERTIES OF CONIC SECTIONS PROVED GEOMETRICALLY. By Rev. H. G. Day, M.A. Part I. The Ellipse, with an ample collection of Problems. Cr. 8vo. 3s. 6d.

*DEAKIN.—RIDER PAPERS ON EUCLID. BOOKS I. AND II. By Rupert Deakin, M.A. 18mo. 1s.

DODGSON.—Works by Charles L. Dodgson, M.A., Student and late Mathematical Lecturer, Christ Church, Oxford.
EUCLID, BOOKS I. AND II. 6th Ed., with words substituted for the Algebraical Symbols used in the 1st Ed. Cr. 8vo. 2s.
EUCLID AND HIS MODERN RIVALS. 2d Ed. Cr. 8vo. 6s.
CURIOSA MATHEMATICA. Part I. A New Theory of Parallels. 3d Ed. Cr. 8vo. 2s.

DREW.—GEOMETRICAL TREATISE ON CONIC SECTIONS. By W. H. Drew, M.A. New Ed., enlarged. Cr. 8vo. 5s.

DUPUIS.—ELEMENTARY SYNTHETIC GEOMETRY OF THE POINT, LINE AND CIRCLE IN THE PLANE. By N. F. Dupuis, M.A., Professor of Pure Mathematics in the University of Queen's College, Kingston, Canada. Gl. 8vo. 4s. 6d.

*HALL and STEVENS.—A TEXT-BOOK OF EUCLID'S ELEMENTS. Including Alternative Proofs, together with additional Theorems and Exercises, classified and arranged. By H. S. Hall, M.A., and F. H. Stevens, M.A., Masters of the Military and Engineering Side, Clifton College. Gl. 8vo. Book I., 1s.; Books I. and II., 1s. 6d.; Books I.-IV., 3s.; Books III.-IV., 2s.; Books III.-VI., 3s.; Books V.-VI. and XI., 2s. 6d.; Books I.-VI. and XI., 4s. 6d.; Book XI., 1s. [KEY. In preparation.

HALSTED.—THE ELEMENTS OF GEOMETRY. By G. B. Halsted, Professor of Pure and Applied Mathematics in the University of Texas. 8vo. 12s. 6d.

HAYWARD.—THE ELEMENTS OF SOLID GEOMETRY. By R. B. Hayward, M.A., F.R.S. Gl. 8vo. 3s.

LOCK.—EUCLID FOR BEGINNERS. Being an Introduction to existing Text-Books. By Rev. J. B. Lock, M.A. [In the Press.

MILNE and DAVIS.—GEOMETRICAL CONICS. Part I. The Parabola. By Rev. J. J. Milne, M.A., and R. F. Davis, M.A. Cr. 8vo. 2s.

*RICHARDSON.—THE PROGRESSIVE EUCLID. Books I. and II. With Notes, Exercises, and Deductions. Edited by A. T. Richardson, M.A., Senior Mathematical Master at the Isle of Wight College. Gl. 8vo. 2s. 6d.

SYLLABUS OF PLANE GEOMETRY (corresponding to Euclid, Books I.-VI.)—Prepared by the Association for the Improvement of Geometrical Teaching. Cr. 8vo. Sewed, 1s.

SYLLABUS OF MODERN PLANE GEOMETRY.—Prepared by the Association for the Improvement of Geometrical Teaching. Cr. 8vo. Sewed, 1s.

*TODHUNTER.—THE ELEMENTS OF EUCLID. By I. Todhunter, F.R.S. 18mo. 3s. 6d. *Books I. and II. 1s. KEY. Cr. 8vo. 6s. 6d.

WILSON.—Works by Ven. Archdeacon Wilson, M.A., formerly Headmaster of Clifton College.
ELEMENTARY GEOMETRY. BOOKS I.-V. Containing the Subjects of Euclid's first Six Books. Following the Syllabus of the Geometrical Association. Ex. fcap. 8vo. 4s. 6d.

GEOMETRICAL DRAWING—TRIGONOMETRY 25

WILSON.—Works by Ven. Archdeacon WILSON—*continued.*
SOLID GEOMETRY AND CONIC SECTIONS. With Appendices on Transversals and Harmonic Division. Ex. fcap. 8vo. 3s. 6d.

GEOMETRICAL DRAWING.

EAGLES.—CONSTRUCTIVE GEOMETRY OF PLANE CURVES. By T. H. EAGLES, M.A., Instructor in Geometrical Drawing and Lecturer in Architecture at the Royal Indian Engineering College, Cooper's Hill. Cr. 8vo. 12s.

EDGAR and PRITCHARD.—NOTE-BOOK ON PRACTICAL SOLID OR DESCRIPTIVE GEOMETRY. Containing Problems with help for Solutions. By J. H. EDGAR and G. S. PRITCHARD. 4th Ed., revised by A. MEEZE. Gl. 8vo. 4s. 6d.

*KITCHENER.—A GEOMETRICAL NOTE-BOOK. Containing Easy Problems in Geometrical Drawing preparatory to the Study of Geometry. For the Use of Schools. By F. E. KITCHENER, M.A., Headmaster of the Newcastle-under-Lyme High School. 4to. 2s.

MILLAR.—ELEMENTS OF DESCRIPTIVE GEOMETRY. By J. B. MILLAR, Civil Engineer, Lecturer on Engineering in the Victoria University, Manchester. 2d Ed. Cr. 8vo. 6s.

PLANT.—PRACTICAL PLANE AND DESCRIPTIVE GEOMETRY. By E. C. PLANT. Globe 8vo.   [*In preparation.*

MENSURATION.

STEVENS.—ELEMENTARY MENSURATION. With Exercises on the Mensuration of Plane and Solid Figures. By F. H. STEVENS, M.A. Gl. 8vo.
[*In preparation.*

TEBAY.—ELEMENTARY MENSURATION FOR SCHOOLS. By S. TEBAY. Ex. fcap. 8vo. 3s. 6d.

*TODHUNTER.—MENSURATION FOR BEGINNERS. By ISAAC TODHUNTER, F.R.S. 18mo. 2s. 6d. KEY. By Rev. FR. L. MCCARTHY. Cr. 8vo. 7s. 6d.

TRIGONOMETRY.

BEASLEY.—AN ELEMENTARY TREATISE ON PLANE TRIGONOMETRY. With Examples. By R. D. BEASLEY, M.A. 9th Ed., revised and enlarged. Cr. 8vo. 3s. 6d.

BOTTOMLEY.—FOUR-FIGURE MATHEMATICAL TABLES. Comprising Logarithmic and Trigonometrical Tables, and Tables of Squares, Square Roots, and Reciprocals. By J. T. BOTTOMLEY, M.A., Lecturer in Natural Philosophy in the University of Glasgow. 8vo. 2s. 6d.

HAYWARD.—THE ALGEBRA OF CO-PLANAR VECTORS AND TRIGONOMETRY. By R. B. HAYWARD, M.A., F.R.S., Assistant Master at Harrow.
[*In the Press.*

JOHNSON.—A TREATISE ON TRIGONOMETRY. By W. E. JOHNSON, M.A., late Scholar and Assistant Mathematical Lecturer at King's College, Cambridge. Cr. 8vo. 8s. 6d.

LEVETT and DAVISON.—ELEMENTS OF TRIGONOMETRY. By RAWDON LEVETT and A. F. DAVISON, Assistant Masters at King Edward's School, Birmingham.   [*In the Press.*

LOCK.—Works by Rev. J. B. LOCK, M.A., Senior Fellow and Bursar of Gonville and Caius College, Cambridge.

*THE TRIGONOMETRY OF ONE ANGLE. Gl. 8vo. 2s. 6d.

*TRIGONOMETRY FOR BEGINNERS, as far as the Solution of Triangles. 3d Ed. Gl. 8vo. 2s. 6d. KEY. Cr. 8vo. 6s. 6d.

*ELEMENTARY TRIGONOMETRY. 6th Ed. (in this edition the chapter on logarithms has been carefully revised). Gl. 8vo. 4s. 6d. KEY. Cr. 8vo. 8s. 6d.

HIGHER TRIGONOMETRY. 5th Ed. Gl. 8vo. 4s. 6d. Both Parts complete in One Volume. Gl. 8vo. 7s. 6d.

M'CLELLAND and PRESTON.—A TREATISE ON SPHERICAL TRIGONOMETRY. With applications to Spherical Geometry and numerous Examples. By W. J. M'CLELLAND, M.A., Principal of the Incorporated Society's School, Santry, Dublin, and T. PRESTON, M.A. Cr. 8vo. 8s. 6d., or: Part I. To the End of Solution of Triangles, 4s. 6d. Part II., 5s.

MATTHEWS.—MANUAL OF LOGARITHMS. By G. F. MATTHEWS, B.A. 8vo. 5s. net.

PALMER.—TEXT-BOOK OF PRACTICAL LOGARITHMS AND TRIGONOMETRY. By J. H. PALMER, Headmaster, R.N., H.M.S. *Cambridge*, Devonport. Gl. 8vo. 4s. 6d.

SNOWBALL.—THE ELEMENTS OF PLANE AND SPHERICAL TRIGONOMETRY. By J. C. SNOWBALL. 14th Ed. Cr. 8vo. 7s. 6d.

TODHUNTER.—Works by ISAAC TODHUNTER, F.R.S.
 *TRIGONOMETRY FOR BEGINNERS. 18mo. 2s. 6d. KEY. Cr. 8vo. 8s. 6d.
 PLANE TRIGONOMETRY. Cr. 8vo. 5s. A New Edition, revised by R. W. HOGG, M.A. Cr. 8vo. 5s. KEY. Cr. 8vo. 10s. 6d.
 A TREATISE ON SPHERICAL TRIGONOMETRY. Cr. 8vo. 4s. 6d.

WOLSTENHOLME.—EXAMPLES FOR PRACTICE IN THE USE OF SEVEN-FIGURE LOGARITHMS. By JOSEPH WOLSTENHOLME, D.Sc., late Professor of Mathematics in the Royal Indian Engineering Coll., Cooper's Hill. 8vo. 5s.

ANALYTICAL GEOMETRY (Plane and Solid).

DYER.—EXERCISES IN ANALYTICAL GEOMETRY. By J. M. DYER, M.A., Assistant Master at Eton. Illustrated. Cr. 8vo. 4s. 6d.

FERRERS.—AN ELEMENTARY TREATISE ON TRILINEAR CO-ORDINATES, the Method of Reciprocal Polars, and the Theory of Projectors. By the Rev. N. M. FERRERS, D.D., F.R.S., Master of Gonville and Caius College, Cambridge. 4th Ed., revised. Cr. 8vo. 6s. 6d.

FROST.—Works by PERCIVAL FROST, D.Sc., F.R.S., Fellow and Mathematical Lecturer at King's College, Cambridge.
 AN ELEMENTARY TREATISE ON CURVE TRACING. 8vo. 12s.
 SOLID GEOMETRY. 3d Ed. Demy 8vo. 16s.
 HINTS FOR THE SOLUTION OF PROBLEMS in the Third Edition of SOLID GEOMETRY. 8vo. 8s. 6d.

JOHNSON.—CURVE TRACING IN CARTESIAN CO-ORDINATES. By W. WOOLSEY JOHNSON, Professor of Mathematics at the U.S. Naval Academy, Annapolis, Maryland. Cr. 8vo. 4s. 6d.

M'CLELLAND.—THE GEOMETRY OF THE CIRCLE. By W. J. M'CLELLAND, M.A. Cr. 8vo. [*In the Press.*

PUCKLE.—AN ELEMENTARY TREATISE ON CONIC SECTIONS AND ALGEBRAIC GEOMETRY. With Numerous Examples and Hints for their Solution. By G. H. PUCKLE, M.A. 5th Ed., revised and enlarged. Cr. 8vo. 7s. 6d.

SMITH.—Works by CHARLES SMITH, M.A., Master of Sidney Sussex College, Cambridge.
 CONIC SECTIONS. 7th Ed. Cr. 8vo. 7s. 6d.
 SOLUTIONS TO CONIC SECTIONS. Cr. 8vo. 10s. 6d.
 AN ELEMENTARY TREATISE ON SOLID GEOMETRY. 2d Ed. Cr. 8vo. 9s. 6d.

TODHUNTER.—Works by ISAAC TODHUNTER, F.R.S.
 PLANE CO-ORDINATE GEOMETRY, as applied to the Straight Line and the Conic Sections. Cr. 8vo. 7s. 6d.
 KEY. By C. W. BOURNE, M.A., Headmaster of King's College School. Cr. 8vo. 10s. 6d.

MATHEMATICS 27

TODHUNTER.—Works by ISAAC TODHUNTER, F.R.S.—*continued.*
EXAMPLES OF ANALYTICAL GEOMETRY OF THREE DIMENSIONS. New Ed., revised. Cr. 8vo. 4s.

PROBLEMS AND QUESTIONS IN MATHEMATICS.

ARMY PRELIMINARY EXAMINATION, 1882-1890, Specimens of Papers set at the. With Answers to the Mathematical Questions. Subjects: Arithmetic, Algebra, Euclid, Geometrical Drawing, Geography, French, English Dictation. Cr. 8vo. 3s. 6d.

CAMBRIDGE SENATE-HOUSE PROBLEMS AND RIDERS, WITH SOLUTIONS:—
1875—PROBLEMS AND RIDERS. By A. G. GREENHILL, F.R.S. Cr. 8vo. 8s. 6d.
1878—SOLUTIONS OF SENATE-HOUSE PROBLEMS. By the Mathematical Moderators and Examiners. Edited by J. W. L. GLAISHER, F.R.S., Fellow of Trinity College, Cambridge. 12s.

CHRISTIE.—A COLLECTION OF ELEMENTARY TEST-QUESTIONS IN PURE AND MIXED MATHEMATICS; with Answers and Appendices on Synthetic Division, and on the Solution of Numerical Equations by Horner's Method. By JAMES R. CHRISTIE, F.R.S. Cr. 8vo. 8s. 6d.

CLIFFORD.—MATHEMATICAL PAPERS. By W. K. CLIFFORD. Edited by R. TUCKER. With an Introduction by H. J. STEPHEN SMITH, M.A. 8vo. 30s.

MILNE.—Works by Rev. JOHN J. MILNE, Private Tutor.
WEEKLY PROBLEM PAPERS. With Notes intended for the use of Students preparing for Mathematical Scholarships, and for Junior Members of the Universities who are reading for Mathematical Honours. Pott 8vo. 4s. 6d.
SOLUTIONS TO WEEKLY PROBLEM PAPERS. Cr. 8vo. 10s. 6d.
COMPANION TO WEEKLY PROBLEM PAPERS. Cr. 8vo. 10s. 6d.

RICHARDSON.—PROGRESSIVE MATHEMATICAL EXERCISES, for Home Work. First Series. By A. T. RICHARDSON, M.A., Senior Mathematical Master at the Isle of Wight College. Gl. 8vo. [*In the Press.*

SANDHURST MATHEMATICAL PAPERS, for admission into the Royal Military College, 1881-1889. Edited by E. J. BROOKSMITH, B.A., Instructor in Mathematics at the Royal Military Academy, Woolwich. Cr. 8vo. 3s. 6d.

WOOLWICH MATHEMATICAL PAPERS, for Admission into the Royal Military Academy, Woolwich, 1880-1890 inclusive. By the same Editor. Cr. 8vo. 6s.

WOLSTENHOLME.—Works by JOSEPH WOLSTENHOLME, D.Sc., late Professor of Mathematics in the Royal Engineering Coll., Cooper's Hill.
MATHEMATICAL PROBLEMS, on Subjects included in the First and Second Divisions of the Schedule of Subjects for the Cambridge Mathematical Tripos Examination. 3d Ed., greatly enlarged. 8vo. 18s.
EXAMPLES FOR PRACTICE IN THE USE OF SEVEN-FIGURE LOGARITHMS. 8vo. 5s.

HIGHER PURE MATHEMATICS.

AIRY.—Works by Sir G. B. AIRY, K.C.B., formerly Astronomer-Royal.
ELEMENTARY TREATISE ON PARTIAL DIFFERENTIAL EQUATIONS. With Diagrams. 2d Ed. Cr. 8vo. 5s. 6d.
ON THE ALGEBRAICAL AND NUMERICAL THEORY OF ERRORS OF OBSERVATIONS AND THE COMBINATION OF OBSERVATIONS. 2d Ed., revised. Cr. 8vo. 6s. 6d.

BOOLE.—THE CALCULUS OF FINITE DIFFERENCES. By G. BOOLE. 3d Ed., revised by J. F. MOULTON, Q.C. Cr. 8vo. 10s. 6d.

EDWARDS.—THE DIFFERENTIAL CALCULUS. By JOSEPH EDWARDS, M.A. With Applications and numerous Examples. Cr. 8vo. 10s. 6d.

FERRERS.—AN ELEMENTARY TREATISE ON SPHERICAL HARMONICS, AND SUBJECTS CONNECTED WITH THEM. By Rev. N. M. FERRERS, D.D., F.R.S., Master of Gonville and Caius College, Cambridge. Cr. 8vo. 7s. 6d.

FORSYTH.—A TREATISE ON DIFFERENTIAL EQUATIONS. By ANDREW RUSSELL FORSYTH, F.R.S., Fellow and Assistant Tutor of Trinity College, Cambridge. 2d Ed. 8vo. 14s.

FROST.—AN ELEMENTARY TREATISE ON CURVE TRACING. By PERCIVAL FROST, M.A., D Sc. 8vo. 12s.

GRAHAM.—GEOMETRY OF POSITION. By R. H. GRAHAM. Cr. 8vo. 7s. 6d.

GREENHILL.—DIFFERENTIAL AND INTEGRAL CALCULUS. By A. G. GREENHILL, Professor of Mathematics to the Senior Class of Artillery Officers, Woolwich. New Ed. Cr. 8vo. 10s. 6d.

APPLICATIONS OF ELLIPTIC FUNCTIONS. By the same. [*In the Press.*

HEMMING.—AN ELEMENTARY TREATISE ON THE DIFFERENTIAL AND INTEGRAL CALCULUS. By G. W. HEMMING, M.A. 2d Ed. 8vo. 9s.

JOHNSON.—Works by WILLIAM WOOLSEY JOHNSON, Professor of Mathematics at the U.S. Naval Academy, Annapolis, Maryland.

INTEGRAL CALCULUS, an Elementary Treatise on the. Founded on the Method of Rates or Fluxions. 8vo. 9s.

CURVE TRACING IN CARTESIAN CO-ORDINATES. Cr. 8vo. 4s. 6d.

A TREATISE ON ORDINARY AND DIFFERENTIAL EQUATIONS. Ex. cr. 8vo. 15s.

KELLAND and TAIT.—INTRODUCTION TO QUATERNIONS, with numerous examples. By P. KELLAND and P. G. TAIT, Professors in the Department of Mathematics in the University of Edinburgh. 2d Ed. Cr. 8vo. 7s. 6d.

KEMPE.—HOW TO DRAW A STRAIGHT LINE: a Lecture on Linkages. By A. B. KEMPE. Illustrated. Cr. 8vo. 1s. 6d.

KNOX.—DIFFERENTIAL CALCULUS FOR BEGINNERS. By ALEXANDER KNOX. Fcap. 8vo. 3s. 6d.

MUIR.—THE THEORY OF DETERMINANTS IN THE HISTORICAL ORDER OF ITS DEVELOPMENT. Part I. Determinants in General. Leibnitz (1693) to Cayley (1841). By THOS. MUIR, Mathematical Master in the High School of Glasgow. 8vo. 10s. 6d.

RICE and JOHNSON.—AN ELEMENTARY TREATISE ON THE DIFFERENTIAL CALCULUS. Founded on the Method of Rates or Fluxions. By J. M. RICE, Professor of Mathematics in the United States Navy, and W. W. JOHNSON, Professor of Mathematics at the United States Naval Academy. 3d Ed., revised and corrected. 8vo. 18s. Abridged Ed. 9s.

TODHUNTER.—Works by ISAAC TODHUNTER, F.R.S.

AN ELEMENTARY TREATISE ON THE THEORY OF EQUATIONS. Cr. 8vo. 7s. 6d.

A TREATISE ON THE DIFFERENTIAL CALCULUS. Cr. 8vo. 10s. 6d. KEY. Cr. 8vo. 10s. 6d.

A TREATISE ON THE INTEGRAL CALCULUS AND ITS APPLICATIONS. Cr. 8vo. 10s. 6d. KEY. Cr. 8vo. 10s. 6d.

A HISTORY OF THE MATHEMATICAL THEORY OF PROBABILITY, from the time of Pascal to that of Laplace. 8vo. 18s.

AN ELEMENTARY TREATISE ON LAPLACE'S, LAMÉ'S, AND BESSEL'S FUNCTIONS. Cr. 8vo. 10s. 6d.

MECHANICS: Statics, Dynamics, Hydrostatics, Hydrodynamics. (See also Physics.)

ALEXANDER and THOMSON.—ELEMENTARY APPLIED MECHANICS. By Prof. T. ALEXANDER and A. W. THOMSON. Part II. Transverse Stress. Cr. 8vo. 10s. 6d.

BALL.—EXPERIMENTAL MECHANICS. A Course of Lectures delivered at the Royal College of Science for Ireland. By Sir R. S. BALL, F.R.S. 2d Ed. Illustrated. Cr. 8vo. 6s.

CLIFFORD.—THE ELEMENTS OF DYNAMIC. An Introduction to the Study of Motion and Rest in Solid and Fluid Bodies. By W. K. CLIFFORD. Part I.— Kinematic. Cr. 8vo. Books I.-III. 7s. 6d.; Book IV. and Appendix, 6s.

MECHANICS

COTTERILL.—APPLIED MECHANICS: An Elementary General Introduction to the Theory of Structures and Machines. By J. H. COTTERILL, F.R.S., Professor of Applied Mechanics in the Royal Naval College, Greenwich. 8vo. 18s.

COTTERILL and SLADE.—LESSONS IN APPLIED MECHANICS. By Prof. J. H. COTTERILL and J. H. SLADE. Fcap. 8vo. 5s. 6d.

DYNAMICS, SYLLABUS OF ELEMENTARY. Part I. Linear Dynamics. With an Appendix on the Meanings of the Symbols in Physical Equations. Prepared by the Association for the Improvement of Geometrical Teaching. 4to. 1s.

GANGUILLET and KUTTER.—A GENERAL FORMULA FOR THE UNIFORM FLOW OF WATER IN RIVERS AND OTHER CHANNELS. By E. GANGUILLET and W. R. KUTTER. Translated, with Additions, including Tables and Diagrams, and the Elements of over 1200 Gaugings of Rivers, Small Channels, and Pipes in English Measure, by R. HERING, Assoc. Am. Soc. C.E., M. Inst. C.E., and J. C. TRAUTWINE Jun., Assoc. Am. Soc. C.E., Assoc. Inst. C.E. 8vo. 17s.

GRAHAM.—GEOMETRY OF POSITION. By R. H. GRAHAM. Cr. 8vo. 7s. 6d.

GREAVES.—Works by JOHN GREAVES, M.A., Fellow and Mathematical Lecturer at Christ's College, Cambridge.
*STATICS FOR BEGINNERS. Gl. 8vo. 3s. 6d.
A TREATISE ON ELEMENTARY STATICS. 2d Ed. Cr. 8vo. 6s. 6d.

GREENHILL.—HYDROSTATICS. By A. G. GREENHILL, Professor of Mathematics to the Senior Class of Artillery Officers, Woolwich. Cr. 8vo. [In preparation.

***HICKS.**—ELEMENTARY DYNAMICS OF PARTICLES AND SOLIDS. By W. M. HICKS, D.Sc., Principal and Professor of Mathematics and Physics, Firth College, Sheffield. Cr. 8vo. 6s. 6d.

JELLETT.—A TREATISE ON THE THEORY OF FRICTION. By JOHN H. JELLETT, B.D., late Provost of Trinity College, Dublin. 8vo. 8s. 6d.

KENNEDY.—THE MECHANICS OF MACHINERY. By A. B. W. KENNEDY, F.R.S. Illustrated. Cr. 8vo. 12s. 6d.

LOCK.—Works by Rev. J. B. LOCK, M.A.
*ELEMENTARY STATICS. 2d Ed. Gl. 8vo. 4s. 6d.
*ELEMENTARY DYNAMICS. 3d Ed. Gl. 8vo. 4s. 6d.
ELEMENTARY HYDROSTATICS. Gl. 8vo. [In preparation.
*MECHANICS FOR BEGINNERS. Gl. 8vo. Part I. MECHANICS OF SOLIDS. 3s. 6d. Part II. MECHANICS OF FLUIDS. [In preparation.

MACGREGOR.—KINEMATICS AND DYNAMICS. An Elementary Treatise. By J. G. MACGREGOR, D.Sc., Munro Professor of Physics in Dalhousie College, Halifax, Nova Scotia. Illustrated. Cr. 8vo. 10s. 6d.

PARKINSON.—AN ELEMENTARY TREATISE ON MECHANICS. By S. PARKINSON, D.D., F.R.S., late Tutor and Prælector of St. John's College, Cambridge. 6th Ed., revised. Cr. 8vo. 9s. 6d.

PIRIE.—LESSONS ON RIGID DYNAMICS. By Rev. G. PIRIE, M.A., Professor of Mathematics in the University of Aberdeen. Cr. 8vo. 6s.

ROUTH.—Works by EDWARD JOHN ROUTH, D.Sc., LL.D., F.R.S., Hon. Fellow of St. Peter's College, Cambridge.
A TREATISE ON THE DYNAMICS OF THE SYSTEM OF RIGID BODIES. With numerous Examples. Two Vols. 8vo. Vol. I.—Elementary Parts. 5th Ed. 14s. Vol. II.—The Advanced Parts. 4th Ed. 14s.
STABILITY OF A GIVEN STATE OF MOTION, PARTICULARLY STEADY MOTION. Adams Prize Essay for 1877. 8vo. 8s. 6d.

***SANDERSON.**—HYDROSTATICS FOR BEGINNERS. By F. W. SANDERSON, M.A., Assistant Master at Dulwich College. Gl. 8vo. 4s. 6d.

TAIT and STEELE.—A TREATISE ON DYNAMICS OF A PARTICLE. By Professor TAIT, M.A., and W. J. STEELE, B.A. 6th Ed., revised. Cr. 8vo. 12s.

TODHUNTER.—Works by ISAAC TODHUNTER, F.R.S.
*MECHANICS FOR BEGINNERS. 18mo. 4s. 6d. KEY. Cr. 8vo. 6s. 6d.
A TREATISE ON ANALYTICAL STATICS. 5th Ed. Edited by Prof. J. D. EVERETT, F.R.S. Cr. 8vo. 10s. 6d.

PHYSICS: Sound, Light, Heat, Electricity, Elasticity, Attractions, etc. (See also Mechanics.)

AIRY.—Works by Sir G. B. AIRY, K.C.B., formerly Astronomer-Royal.
  ON SOUND AND ATMOSPHERIC VIBRATIONS. With the Mathematical Elements of Music. 2d Ed., revised and enlarged. Cr. 8vo. 9s.
  GRAVITATION: An Elementary Explanation of the Principal Perturbations in the Solar System. 2d Ed. Cr. 8vo. 7s. 6d.

CLAUSIUS.—MECHANICAL THEORY OF HEAT. By R. CLAUSIUS. Translated by W. R. BROWNE, M.A. Cr. 8vo. 10s. 6d.

CUMMING.—AN INTRODUCTION TO THE THEORY OF ELECTRICITY. By LINNÆUS CUMMING, M.A., Assistant Master at Rugby. Illustrated. Cr. 8vo. 8s. 6d.

DANIELL.—A TEXT-BOOK OF THE PRINCIPLES OF PHYSICS. By ALFRED DANIELL, D.Sc. Illustrated. 2d Ed., revised and enlarged. 8vo. 21s.

DAY.—ELECTRIC LIGHT ARITHMETIC. By R. E. DAY, Evening Lecturer in Experimental Physics at King's College, London. Pott 8vo. 2s.

EVERETT.—ILLUSTRATIONS OF THE C. G. S. SYSTEM OF UNITS WITH TABLES OF PHYSICAL CONSTANTS. By J. D. EVERETT, F.R.S., Professor of Natural Philosophy, Queen's College, Belfast. New Ed. Ex. fcap. 8vo. 5s.

FERRERS.—AN ELEMENTARY TREATISE ON SPHERICAL HARMONICS, and Subjects connected with them. By Rev. N. M. FERRERS, D.D., F.R.S., Master of Gonville and Caius College, Cambridge. Cr. 8vo. 7s. 6d.

FESSENDEN.—PHYSICS FOR PUBLIC SCHOOLS. By C. FESSENDEN. Illustrated. Fcap. 8vo. [*In the Press.*

GRAY.—THE THEORY AND PRACTICE OF ABSOLUTE MEASUREMENTS IN ELECTRICITY AND MAGNETISM. By A. GRAY, F.R.S.E., Professor of Physics in the University College of North Wales. Two Vols. Cr. 8vo. Vol. I. 12s. 6d. [Vol. II. *In the Press.*
  ABSOLUTE MEASUREMENTS IN ELECTRICITY AND MAGNETISM. 2d Ed., revised and greatly enlarged. Fcap. 8vo. 5s. 6d.

IBBETSON.—THE MATHEMATICAL THEORY OF PERFECTLY ELASTIC SOLIDS, with a Short Account of Viscous Fluids. By W. J. IBBETSON, late Senior Scholar of Clare College, Cambridge. 8vo. 21s.

JOHNSON.—NATURE'S STORY BOOKS. I. Sunshine. By AMY JOHNSON, LL.A. Illustrated. [*In the Press.*

*****JONES.**—EXAMPLES IN PHYSICS. Containing over 1000 Problems with Answers and numerous solved Examples. Suitable for candidates preparing for the Intermediate, Science, Preliminary, Scientific, and other Examinations of the University of London. By Prof. D. E. JONES, B.Sc., Fcap. 8vo. 3s. 6d.

*ELEMENTARY LESSONS IN HEAT, LIGHT, AND SOUND. By the same. Gl. 8vo. 2s. 6d.

LOCKYER.—CONTRIBUTIONS TO SOLAR PHYSICS. By J. NORMAN LOCKYER, F.R.S. With Illustrations. Royal 8vo. 31s. 6d.

LODGE.—MODERN VIEWS OF ELECTRICITY. By OLIVER J. LODGE, F.R.S., Professor of Experimental Physics in University College, Liverpool. Illustrated. Cr. 8vo. 6s. 6d.

LOEWY.—*QUESTIONS AND EXAMPLES ON EXPERIMENTAL PHYSICS: Sound, Light, Heat, Electricity, and Magnetism. By B. LOEWY, Examiner in Experimental Physics to the College of Preceptors. Fcap. 8vo. 2s.
  *A GRADUATED COURSE OF NATURAL SCIENCE FOR ELEMENTARY AND TECHNICAL SCHOOLS AND COLLEGES. By the same. Gl. 8vo. In Three Parts. Part I. FIRST YEAR'S COURSE. 2s. Part II. SECOND YEAR'S COURSE. 2s. 6d.

LUPTON.—NUMERICAL TABLES AND CONSTANTS IN ELEMENTARY SCIENCE. By S. LUPTON, M.A., late Assistant Master at Harrow. Ex. fcap. 8vo. 2s. 6d.

PHYSICS 31

MACFARLANE.—PHYSICAL ARITHMETIC. By A. MACFARLANE, D.Sc., late Examiner in Mathematics at the University of Edinburgh. Cr. 8vo. 7s. 6d.

*****MAYER.**—SOUND: A Series of Simple, Entertaining, and Inexpensive Experiments in the Phenomena of Sound. By A. M. MAYER, Professor of Physics in the Stevens Institute of Technology. Illustrated. Cr. 8vo. 3s. 6d.

*****MAYER and BARNARD.**—LIGHT: A Series of Simple, Entertaining, and Inexpensive Experiments in the Phenomena of Light. By A. M. MAYER and C. BARNARD. Illustrated. Cr. 8vo. 2s. 6d.

MOLLOY.—GLEANINGS IN SCIENCE: Popular Lectures on Scientific Subjects. By the Rev. GERALD MOLLOY, D.Sc., Rector of the Catholic University of Ireland. 8vo. 7s. 6d.

NEWTON.—PRINCIPIA. Edited by Prof. Sir W. THOMSON, P.R.S., and Prof. BLACKBURNE. 4to. 31s. 6d.

THE FIRST THREE SECTIONS OF NEWTON'S PRINCIPIA. With Notes and Illustrations. Also a Collection of Problems, principally intended as Examples of Newton's Methods. By P. FROST, M.A., D.Sc. 3d Ed. 8vo. 12s.

PARKINSON.—A TREATISE ON OPTICS. By S. PARKINSON, D.D., F.R.S., late Tutor and Prælector of St. John's College, Cambridge. 4th Ed., revised and enlarged. Cr. 8vo. 10s. 6d.

PEABODY.—THERMODYNAMICS OF THE STEAM-ENGINE AND OTHER HEAT-ENGINES. By CECIL H. PEABODY, Associate Professor of Steam Engineering, Massachusetts Institute of Technology. 8vo. 21s.

PERRY.—STEAM: An Elementary Treatise. By JOHN PERRY, Professor of Mechanical Engineering and Applied Mechanics at the Technical College, Finsbury. 18mo. 4s. 6d.

PICKERING.—ELEMENTS OF PHYSICAL MANIPULATION. By Prof. EDWARD C. PICKERING. Medium 8vo. Part I., 12s. 6d. Part II., 14s.

PRESTON.—THE THEORY OF LIGHT. By THOMAS PRESTON, M.A. Illustrated. 8vo. 12s. 6d.

THE THEORY OF HEAT. By the same Author. 8vo. [*In preparation.*

RAYLEIGH.—THE THEORY OF SOUND. By Lord RAYLEIGH, F.R.S. 8vo. Vol. I., 12s. 6d. Vol. II., 12s. 6d. [Vol. III. *In the Press.*

SHANN.—AN ELEMENTARY TREATISE ON HEAT, IN RELATION TO STEAM AND THE STEAM-ENGINE. By G. SHANN, M.A. Illustrated. Cr. 8vo. 4s. 6d.

SPOTTISWOODE.—POLARISATION OF LIGHT. By the late W. SPOTTISWOODE, F.R.S. Illustrated. Cr. 8vo. 3s. 6d.

STEWART.—Works by BALFOUR STEWART, F.R.S., late Langworthy Professor of Physics in the Owens College, Victoria University, Manchester.

*PRIMER OF PHYSICS. Illustrated. With Questions. 18mo. 1s.
*LESSONS IN ELEMENTARY PHYSICS. Illustrated. Fcap. 8vo. 4s. 6d.
*QUESTIONS. By Prof. T. H. CORE. Fcap. 8vo. 2s.

STEWART and GEE.—LESSONS IN ELEMENTARY PRACTICAL PHYSICS. By BALFOUR STEWART, F.R.S., and W. W. HALDANE GEE, B.Sc. Cr. 8vo. Vol. I. GENERAL PHYSICAL PROCESSES. 6s. Vol. II. ELECTRICITY AND MAGNETISM. 7s. 6d. [Vol. III. OPTICS, HEAT, AND SOUND. *In the Press.*

*PRACTICAL PHYSICS FOR SCHOOLS AND THE JUNIOR STUDENTS OF COLLEGES. Gl. 8vo. Vol. I. ELECTRICITY AND MAGNETISM. 2s. 6d. [Vol. II. OPTICS, HEAT, AND SOUND. *In the Press.*

STOKES.—ON LIGHT. Burnett Lectures, delivered in Aberdeen in 1883-4-5. By Sir G. G. STOKES, F.R.S., Lucasian Professor of Mathematics in the University of Cambridge. First Course: ON THE NATURE OF LIGHT. Second Course: ON LIGHT AS A MEANS OF INVESTIGATION. Third Course: ON THE BENEFICIAL EFFECTS OF LIGHT. Cr. 8vo. 7s. 6d.
⁎ The 2d and 3d Courses may be had separately. Cr. 8vo. 2s. 6d. each.

STONE.—AN ELEMENTARY TREATISE ON SOUND. By W. H. STONE. Illustrated. Fcap. 8vo. 3s. 6d.

TAIT.—HEAT. By P. G. TAIT, Professor of Natural Philosophy in the University of Edinburgh. Cr. 8vo. 6s.

MATHEMATICS

LECTURES ON SOME RECENT ADVANCES IN PHYSICAL SCIENCE. By the same. 3d Edition. Crown 8vo. 9s.

TAYLOR.—SOUND AND MUSIC. An Elementary Treatise on the Physical Constitution of Musical Sounds and Harmony, including the Chief Acoustical Discoveries of Professor Helmholtz. By SEDLEY TAYLOR, M.A. Illustrated. 2d Ed. Ex. cr. 8vo. 8s. 6d.

*THOMPSON. — ELEMENTARY LESSONS IN ELECTRICITY AND MAGNETISM. By SILVANUS P. THOMPSON, Principal and Professor of Physics in the Technical College, Finsbury. Illustrated. New Ed., revised. Fcap. 8vo. 4s. 6d.

THOMSON.—Works by J. J. THOMSON, Professor of Experimental Physics in the University of Cambridge.

A TREATISE ON THE MOTION OF VORTEX RINGS. Adams Prize Essay. 1882. 8vo. 6s.

APPLICATIONS OF DYNAMICS TO PHYSICS AND CHEMISTRY. Cr. 8vo. 7s. 6d.

THOMSON.—Works by Sir W. THOMSON, P.R.S., Professor of Natural Philosophy in the University of Glasgow.

ELECTROSTATICS AND MAGNETISM, REPRINTS OF PAPERS ON. 2d Ed. 8vo. 18s.

POPULAR LECTURES AND ADDRESSES. 3 Vols. Illustrated. Cr. 8vo. Vol. I. CONSTITUTION OF MATTER. 7s. 6d. Vol. III. NAVIGATION. 7s. 6d.

TODHUNTER.—Works by ISAAC TODHUNTER, F.R.S.

AN ELEMENTARY TREATISE ON LAPLACE'S, LAME'S, AND BESSEL'S FUNCTIONS. Crown 8vo. 10s. 6d.

A HISTORY OF THE MATHEMATICAL THEORIES OF ATTRACTION, AND THE FIGURE OF THE EARTH, from the time of Newton to that of Laplace. 2 vols. 8vo. 24s.

TURNER.—A COLLECTION OF EXAMPLES ON HEAT AND ELECTRICITY. By H. H. TURNER, Fellow of Trinity College, Cambridge. Cr. 8vo. 2s. 6d.

WRIGHT.—LIGHT: A Course of Experimental Optics, chiefly with the Lantern. By LEWIS WRIGHT. Illustrated. Cr. 8vo. 7s. 6d.

ASTRONOMY.

AIRY.—Works by Sir G. B. AIRY, K.C.B., formerly Astronomer-Royal.

*POPULAR ASTRONOMY. 7th Ed. Revised by H. H. TURNER, M.A. 18mo. 4s. 6d.

GRAVITATION: An Elementary Explanation of the Principal Perturbations in the Solar System. 2d Ed. Cr. 8vo. 7s. 6d.

CHEYNE.—AN ELEMENTARY TREATISE ON THE PLANETARY THEORY. By C. H. H. CHEYNE. With Problems. 3d Ed. Edited by Rev. A. FREEMAN, M.A., F.R.A.S. Cr. 8vo. 7s. 6d.

CLARK and SADLER.—THE STAR GUIDE. By L. CLARK and H. SADLER. Roy. 8vo. 5s.

CROSSLEY, GLEDHILL, and WILSON.—A HANDBOOK OF DOUBLE STARS. By E. CROSSLEY, J. GLEDHILL, and J. M. WILSON. 8vo. 21s.

CORRECTIONS TO THE HANDBOOK OF DOUBLE STARS. 8vo. 1s.

FORBES.—TRANSIT OF VENUS. By G. FORBES, Professor of Natural Philosophy in the Andersonian University, Glasgow. Illustrated. Cr. 8vo. 3s. 6d.

GODFRAY.—Works by HUGH GODFRAY, M.A., Mathematical Lecturer at Pembroke College, Cambridge.

A TREATISE ON ASTRONOMY. 4th Ed. 8vo. 12s. 6d.

AN ELEMENTARY TREATISE ON THE LUNAR THEORY, with a brief Sketch of the Problem up to the time of Newton. 2d Ed., revised. Cr. 8vo. 5s. 6d.

LOCKYER.—Works by J. NORMAN LOCKYER, F.R.S.

*PRIMER OF ASTRONOMY. Illustrated. 18mo. 1s.

*ELEMENTARY LESSONS IN ASTRONOMY. With Spectra of the Sun, Stars, and Nebulæ, and numerous Illustrations. 36th Thousand. Revised throughout. Fcap. 8vo. 5s. 6d.

ASTRONOMY—HISTORICAL

*QUESTIONS ON LOCKYER'S ELEMENTARY LESSONS IN ASTRONOMY. By J. FORBES ROBERTSON. 18mo. 1s. 6d.
THE CHEMISTRY OF THE SUN. Illustrated. 8vo. 14s.
THE METEORITIC HYPOTHESIS OF THE ORIGIN OF COSMICAL SYSTEMS. Illustrated. 8vo. 17s. net.
THE EVOLUTION OF THE HEAVENS AND THE EARTH. Cr. 8vo. Illustrated. [*In the Press.*
LOCKYER and SEABROKE.—STAR-GAZING PAST AND PRESENT. By J. NORMAN LOCKYER, F.R.S. Expanded from Shorthand Notes with the assistance of G. M. SEABROKE, F.R.A.S. Royal 8vo. 21s.
NEWCOMB.—POPULAR ASTRONOMY. By S. NEWCOMB, LL.D., Professor U.S. Naval Observatory. Illustrated. 2d Ed., revised. 8vo. 18s.

HISTORICAL.

BALL.—A SHORT ACCOUNT OF THE HISTORY OF MATHEMATICS. By W. W. R. BALL, M.A. Cr. 8vo. 10s. 6d.

NATURAL SCIENCES.

Chemistry; Physical Geography, Geology, and Mineralogy; Biology; Medicine.

(For MECHANICS, PHYSICS, and ASTRONOMY, see *MATHEMATICS*.)

CHEMISTRY.

ARMSTRONG.—A MANUAL OF INORGANIC CHEMISTRY. By HENRY ARMSTRONG, F.R.S., Professor of Chemistry in the City and Guilds of London Technical Institute. Cr. 8vo. [*In preparation.*
*COHEN.—THE OWENS COLLEGE COURSE OF PRACTICAL ORGANIC CHEMISTRY. By JULIUS B. COHEN, Ph.D., Assistant Lecturer on Chemistry in the Owens College, Manchester. With a Preface by Sir HENRY ROSCOE, F.R.S., and C. SCHORLEMMER, F.R.S. Fcap. 8vo. 2s. 6d.
COOKE.—ELEMENTS OF CHEMICAL PHYSICS. By JOSIAH P. COOKE, Jun., Erving Professor of Chemistry and Mineralogy in Harvard University. 4th Ed. 8vo. 21s.
FLEISCHER.—A SYSTEM OF VOLUMETRIC ANALYSIS. By EMIL FLEISCHER. Translated, with Notes and Additions, by M. M. P. MUIR, F.R.S.E. Illustrated. Cr. 8vo. 7s. 6d.
FRANKLAND.—A HANDBOOK OF AGRICULTURAL CHEMICAL ANALYSIS. By P. F. FRANKLAND, F.R.S., Professor of Chemistry in University College, Dundee. Cr. 8vo. 7s. 6d.
HARTLEY.—A COURSE OF QUANTITATIVE ANALYSIS FOR STUDENTS. By W. NOEL HARTLEY, F.R.S., Professor of Chemistry and of Applied Chemistry, Science and Art Department, Royal College of Science, Dublin. Gl. 8vo. 5s.
HEMPEL.—METHODS OF GAS ANALYSIS. By Dr. WALTHER HEMPEL. Translated by Dr. L. M. DENNIS. [*In preparation.*
HIORNS.—PRACTICAL METALLURGY AND ASSAYING. A Text-Book for the use of Teachers, Students, and Assayers. By ARTHUR H. HIORNS, Principal of the School of Metallurgy, Birmingham and Midland Institute. Illustrated. Gl. 8vo. 6s.
A TEXT-BOOK OF ELEMENTARY METALLURGY FOR THE USE OF STUDENTS. To which is added an Appendix of Examination Questions, embracing the whole of the Questions set in the three stages of the subject by the Science and Art Department for the past twenty years. By the same. Gl. 8vo. 4s.

IRON AND STEEL MANUFACTURE. A Text-Book for Beginners. By the same. Illustrated. Gl. 8vo. 3s. 6d.

MIXED METALS OR METALLIC ALLOYS. By the same. Gl. 8vo. 6s.

JONES.—*THE OWENS COLLEGE JUNIOR COURSE OF PRACTICAL CHEMISTRY. By FRANCIS JONES, F.R.S.E., Chemical Master at the Grammar School, Manchester. With Preface by Sir HENRY ROSCOE, F.R.S. Illustrated. Fcap. 8vo. 2s. 6d.

*QUESTIONS ON CHEMISTRY. A Series of Problems and Exercises in Inorganic and Organic Chemistry. By the same. Fcap. 8vo. 3s.

LANDAUER.—BLOWPIPE ANALYSIS. By J. LANDAUER. Authorised English Edition by J. TAYLOR and W. E. KAY, of Owens College, Manchester.
[*New Edition in the Press.*

LOCKYER.—THE CHEMISTRY OF THE SUN. By J. NORMAN LOCKYER, F.R.S. Illustrated. 8vo. 14s.

LUPTON.—CHEMICAL ARITHMETIC. With 1200 Problems. By S. LUPTON, M.A. 2d Ed., revised and abridged. Fcap. 8vo. 4s. 6d.

MANSFIELD.—A THEORY OF SALTS. By C. B. MANSFIELD. Crown 8vo. 14s.

MELDOLA.—THE CHEMISTRY OF PHOTOGRAPHY. By RAPHAEL MELDOLA, F.R.S., Professor of Chemistry in the Technical College, Finsbury. Cr. 8vo. 6s.

MEYER.—HISTORY OF CHEMISTRY FROM THE EARLIEST TIMES TO THE PRESENT DAY. By ERNST VON MEYER, Ph.D. Translated by GEORGE McGOWAN, Ph.D. 8vo. 14s. net.

MIXTER.—AN ELEMENTARY TEXT-BOOK OF CHEMISTRY. By WILLIAM G. MIXTER, Professor of Chemistry in the Sheffield Scientific School of Yale College. 2d and revised Ed. Cr. 8vo. 7s. 6d.

MUIR.—PRACTICAL CHEMISTRY FOR MEDICAL STUDENTS. Specially arranged for the first M.B. Course. By M. M. P. MUIR, F.R.S.E., Fellow and Prælector in Chemistry at Gonville and Caius College, Cambridge. Fcap. 8vo. 1s. 6d.

MUIR and WILSON.—THE ELEMENTS OF THERMAL CHEMISTRY. By M. M. P. MUIR, F.R.S.E.; assisted by D. M. WILSON. 8vo. 12s. 6d.

OSTWALD.—OUTLINES OF GENERAL CHEMISTRY (PHYSICAL AND THEORETICAL). By Prof. W. OSTWALD. Translated by JAMES WALKER, D.Sc., Ph.D. 8vo. 10s. net.

RAMSAY.—EXPERIMENTAL PROOFS OF CHEMICAL THEORY FOR BEGINNERS. By WILLIAM RAMSAY, F.R.S., Professor of Chemistry in University College, London. Pott 8vo. 2s. 6d.

REMSEN.—Works by IRA REMSEN, Professor of Chemistry in the Johns Hopkins University, U.S.A.

COMPOUNDS OF CARBON: or, Organic Chemistry, an Introduction to the Study of. Cr. 8vo. 6s. 6d.

AN INTRODUCTION TO THE STUDY OF CHEMISTRY (INORGANIC CHEMISTRY). Cr. 8vo. 6s. 6d.

*THE ELEMENTS OF CHEMISTRY. A Text-Book for Beginners. Fcap. 8vo. 2s. 6d.

A TEXT-BOOK OF INORGANIC CHEMISTRY. 8vo. 16s.

ROSCOE.—Works by Sir HENRY E. ROSCOE, F.R.S., formerly Professor of Chemistry in the Owens College, Victoria University, Manchester.

*PRIMER OF CHEMISTRY. Illustrated. With Questions. 18mo. 1s.

*LESSONS IN ELEMENTARY CHEMISTRY, INORGANIC AND ORGANIC. With Illustrations and Chromolitho of the Solar Spectrum, and of the Alkalies and Alkaline Earths. Fcap. 8vo. 4s. 6d.

ROSCOE and SCHORLEMMER.—INORGANIC AND ORGANIC CHEMISTRY. A Complete Treatise on Inorganic and Organic Chemistry. By Sir HENRY E. ROSCOE, F.R.S., and Prof. C. SCHORLEMMER, F.R.S. Illustrated. 8vo.

Vols. I. and II. INORGANIC CHEMISTRY. Vol. I.—The Non-Metallic Elements. 2d Ed. 21s. Vol. II. Part I.—Metals. 18s. Part II.—Metals. 18s.

Vol. III.—ORGANIC CHEMISTRY. THE CHEMISTRY OF THE HYDROCARBONS and their Derivatives. Five Parts. Parts I., II. and IV 21s. Parts III. and V. 18s. each.

ROSCOE and SCHUSTER.—SPECTRUM ANALYSIS. Lectures delivered in 1868. By Sir HENRY ROSCOE, F.R.S. 4th Ed., revised and considerably enlarged by the Author and by A. SCHUSTER, F.R.S., Ph.D., Professor of Applied Mathematics in the Owens College, Victoria University. With Appendices, Illustrations, and Plates. 8vo. 21s.

*THORPE.—A SERIES OF CHEMICAL PROBLEMS. With Key. For use in Colleges and Schools. By T. E. THORPE, B.Sc. (Vic.), Ph.D., F.R.S. Revised and Enlarged by W. TATE, Assoc.N.S.S. With Preface by Sir H. E. ROSCOE, F.R.S. New Ed. Fcap. 8vo. 2s.

THORPE and RÜCKER.—A TREATISE ON CHEMICAL PHYSICS. By Prof. T. E. THORPE, F.R.S., and Prof. A. W. RÜCKER, F.R.S. Illustrated. 8vo.
[*In preparation.*

WURTZ.—A HISTORY OF CHEMICAL THEORY. By AD. WURTZ. Translated by HENRY WATTS, F.R.S. Crown 8vo. 6s.

PHYSICAL GEOGRAPHY, GEOLOGY, AND MINERALOGY.

BLANFORD.—THE RUDIMENTS OF PHYSICAL GEOGRAPHY FOR THE USE OF INDIAN SCHOOLS; with a Glossary of Technical Terms employed. By H. F. BLANFORD, F.G.S. Illustrated. Cr. 8vo. 2s. 6d.

FERREL.—A POPULAR TREATISE ON THE WINDS. Comprising the General Motions of the Atmosphere, Monsoons, Cyclones, Tornadoes, Waterspouts, Hailstorms, etc. By WILLIAM FERREL, M.A., Member of the American National Academy of Sciences. 8vo. 18s.

FISHER.—PHYSICS OF THE EARTH'S CRUST. By the Rev. OSMOND FISHER, M.A., F.G.S., Hon. Fellow of King's College, London. 2d Ed., altered and enlarged. 8vo. 12s.

GEIKIE.—Works by Sir ARCHIBALD GEIKIE, F.R.S., Director-General of the Geological Survey of the United Kingdom.
*PRIMER OF PHYSICAL GEOGRAPHY. Illustrated. With Questions. 18mo. 1s.
*ELEMENTARY LESSONS IN PHYSICAL GEOGRAPHY. Illustrated. Fcap. 8vo. 4s. 6d. *QUESTIONS ON THE SAME. 1s. 6d.
*PRIMER OF GEOLOGY. Illustrated. 18mo. 1s.
*CLASS-BOOK OF GEOLOGY. Illustrated. New and Cheaper Ed. Cr. 8vo. 4s. 6d.
TEXT-BOOK OF GEOLOGY. Illustrated. 2d Ed., 7th Thousand, revised and enlarged. 8vo. 28s.
OUTLINES OF FIELD GEOLOGY. Illustrated. New Ed., revised and enlarged. Gl. 8vo. 3s. 6d.
THE SCENERY AND GEOLOGY OF SCOTLAND, VIEWED IN CONNEXION WITH ITS PHYSICAL GEOLOGY. Illustrated. Cr. 8vo. 12s. 6d.

HUXLEY.—PHYSIOGRAPHY. An Introduction to the Study of Nature. By T. H. HUXLEY, F.R.S. Illustrated. New and Cheaper Edition. Cr. 8vo. 6s.

LOCKYER.—OUTLINES OF PHYSIOGRAPHY—THE MOVEMENTS OF THE EARTH. By J. NORMAN LOCKYER, F.R.S., Examiner in Physiography for the Science and Art Department. Illustrated. Cr. 8vo. Sewed, 1s. 6d.

MIERS.—A TREATISE ON MINERALOGY. By H. A. MIERS, of the British Museum. 8vo.
[*In preparation.*

PHILLIPS. A TREATISE ON ORE DEPOSITS. By J. ARTHUR PHILLIPS, F.R.S. Illustrated. 8vo. 25s.

ROSENBUSCH and IDDINGS.—MICROSCOPICAL PHYSIOGRAPHY OF THE ROCK-MAKING MINERALS: AN AID TO THE MICROSCOPICAL STUDY OF ROCKS. By H. ROSENBUSCH. Translated and Abridged by J. P. IDDINGS. Illustrated. 8vo. 24s.

WILLIAMS.—ELEMENTS OF CRYSTALLOGRAPHY FOR STUDENTS OF CHEMISTRY, PHYSICS, AND MINERALOGY. By G. H. WILLIAMS, Ph.D. Cr. 8vo. 6s.

BIOLOGY.

ALLEN.—ON THE COLOURS OF FLOWERS, as Illustrated in the British Flora. By GRANT ALLEN. Illustrated. Cr. 8vo. 3s. 6d.

BALFOUR.—A TREATISE ON COMPARATIVE EMBRYOLOGY. By F. M. BALFOUR, F.R.S., Fellow and Lecturer of Trinity College, Cambridge. Illustrated. 2d Ed., reprinted without alteration from the 1st Ed. 2 vols. 8vo. Vol. I. 18s. Vol. II. 21s.

BALFOUR and WARD.—A GENERAL TEXT-BOOK OF BOTANY. By ISAAC BAYLEY BALFOUR, F.R.S., Professor of Botany in the University of Edinburgh, and H. MARSHALL WARD, F.R.S., Professor of Botany in the Royal Indian Engineering College, Cooper's Hill. 8vo. [*In preparation.*

BATESON.—MATERIALS FOR THE STUDY OF VARIATION IN ANIMALS. Part I. Discontinuous Variation. By W. BATESON, M.A., Balfour Student and Fellow of St. John's College, Cambridge. 8vo. Illustrated. [*In the Press.*

*****BETTANY.**—FIRST LESSONS IN PRACTICAL BOTANY. By G. T. BETTANY. 18mo. 1s.

*****BOWER.**—A COURSE OF PRACTICAL INSTRUCTION IN BOTANY. By F. O. BOWER, D.Sc., F.R.S., Regius Professor of Botany in the University of Glasgow. New Ed., revised. Cr. 8vo. 10s. 6d. Abridged Ed. [*In preparation.*

BUCKTON.—MONOGRAPH OF THE BRITISH CICADÆ, OR TETTIGIDÆ. By G. B. BUCKTON. In 8 parts, Quarterly. Part I. January, 1890. 8vo. Parts I.-VI. ready. 8s. each, net. Vol. I. 33s. 6d. net.

CHURCH and SCOTT.—MANUAL OF VEGETABLE PHYSIOLOGY. By Professor A. H. CHURCH, and D. H. SCOTT, D.Sc., Lecturer in the Normal School of Science. Illustrated. Cr. 8vo. [*In preparation.*

COPE.—THE ORIGIN OF THE FITTEST. Essays on Evolution. By E. D. COPE, M.A., Ph.D. 8vo. 12s. 6d.

COUES.—HANDBOOK OF FIELD AND GENERAL ORNITHOLOGY. By Prof. ELLIOTT COUES, M.A. Illustrated. 8vo. 10s. net.

DARWIN.—MEMORIAL NOTICES OF CHARLES DARWIN, F.R.S., etc. By T. H. HUXLEY, F.R.S., G. J. ROMANES, F.R.S., ARCHIBALD GEIKIE, F.R.S., and W. THISELTON DYER, F.R.S. Reprinted from *Nature.* With a Portrait. Cr. 8vo. 2s. 6d.

EIMER.—ORGANIC EVOLUTION AS THE RESULT OF THE INHERITANCE OF ACQUIRED CHARACTERS ACCORDING TO THE LAWS OF ORGANIC GROWTH. By Dr. G. H. THEODOR EIMER. Translated by J. T. CUNNINGHAM, F.R.S.E., late Fellow of University College, Oxford. 8vo. 12s. 6d.

FEARNLEY.—A MANUAL OF ELEMENTARY PRACTICAL HISTOLOGY. By WILLIAM FEARNLEY. Illustrated. Cr. 8vo. 7s. 6d.

FLOWER and GADOW.—AN INTRODUCTION TO THE OSTEOLOGY OF THE MAMMALIA. By W. H. FLOWER, F.R.S., Director of the Natural History Departments of the British Museum. Illustrated. 3d Ed. Revised with the assistance of HANS GADOW, Ph.D., Lecturer on the Advanced Morphology of Vertebrates in the University of Cambridge. Cr. 8vo. 10s. 6d.

FOSTER.—Works by MICHAEL FOSTER, M.D., F.R.S., Professor of Physiology in the University of Cambridge.

  *PRIMER OF PHYSIOLOGY. Illustrated. 18mo. 1s.

  A TEXT-BOOK OF PHYSIOLOGY. Illustrated. 5th Ed., largely revised. 8vo. Part I., comprising Book I. Blood—The Tissues of Movement, The Vascular Mechanism. 10s. 6d. Part II., comprising Book II. The Tissues of Chemical Action, with their Respective Mechanisms—Nutrition. 10s. 6d. Part III. The Central Nervous System. 7s. 6d. Part IV., comprising the remainder of Book III. The Senses and Some Special Muscular Mechanisms; and Book IV. The Tissues and Mechanisms of Reproduction.

FOSTER and BALFOUR.—THE ELEMENTS OF EMBRYOLOGY. By Prof. MICHAEL FOSTER, M.D., F.R.S., and the late F. M. BALFOUR, F.R.S., Professor of Animal Morphology in the University of Cambridge. 2d Ed., revised. Edited by A. SEDGWICK, M.A., Fellow and Assistant Lecturer of Trinity College,

Cambridge, and W. HEAPE, M.A., late Demonstrator in the Morphological Laboratory of the University of Cambridge. Illustrated. Cr. 8vo. 10s. 6d.

FOSTER and LANGLEY.—A COURSE OF ELEMENTARY PRACTICAL PHYSIOLOGY AND HISTOLOGY. By Prof. MICHAEL FOSTER, M.D., F.R.S., and J. N. LANGLEY, F.R.S., Fellow of Trinity College, Cambridge. 6th Ed. Cr. 8vo. 7s. 6d.

GAMGEE.—A TEXT-BOOK OF THE PHYSIOLOGICAL CHEMISTRY OF THE ANIMAL BODY. Including an Account of the Chemical Changes occurring in Disease. By A. GAMGEE, M.D., F.R.S. Illustrated. 8vo. Vol. I. 18s.

GOODALE.—PHYSIOLOGICAL BOTANY. I. Outlines of the Histology of Phænogamous Plants. II. Vegetable Physiology. By GEORGE LINCOLN GOODALE, M.A., M.D., Professor of Botany in Harvard University. 8vo. 10s. 6d.

GRAY.—STRUCTURAL BOTANY, OR ORGANOGRAPHY ON THE BASIS OF MORPHOLOGY. To which are added the Principles of Taxonomy and Phytography, and a Glossary of Botanical Terms. By Prof. ASA GRAY, LL.D. 8vo. 10s. 6d.

THE SCIENTIFIC PAPERS OF ASA GRAY. Selected by C. SPRAGUE SARGENT. 2 vols. Vol. I. Reviews of Works on Botany and Related Subjects, 1834-1887. Vol. II. Essays, Biographical Sketches, 1841-1886. 8vo. 21s.

HAMILTON.—A SYSTEMATIC AND PRACTICAL TEXT-BOOK OF PATHOLOGY. By D. J. HAMILTON, F.R.S.E., Professor of Pathological Anatomy in the University of Aberdeen. Illustrated. 8vo. Vol. I. 25s.

HARTIG.—TEXT-BOOK OF THE DISEASES OF TREES. By Dr. ROBERT HARTIG. Translated by WM. SOMERVILLE, B.Sc., D.Œ., Professor of Agriculture and Forestry, Durham College of Science, Newcastle-on-Tyne. Edited, with Introduction, by Prof. H. MARSHALL WARD. 8vo. [In preparation.

HOOKER.—Works by Sir JOSEPH HOOKER, F.R.S., &c.
*PRIMER OF BOTANY. Illustrated. 18mo. 1s.
THE STUDENT'S FLORA OF THE BRITISH ISLANDS. 3d Ed., revised. Gl. 8vo. 10s. 6d.

HOWES.—AN ATLAS OF PRACTICAL ELEMENTARY BIOLOGY. By G. B. HOWES, Assistant Professor of Zoology, Normal School of Science and Royal School of Mines. With a Preface by Prof. T. H. HUXLEY, F.R.S. 4to. 14s.

HUXLEY.—Works by Prof. T. H. HUXLEY, F.R.S.
*INTRODUCTORY PRIMER OF SCIENCE. 18mo. 1s.
*LESSONS IN ELEMENTARY PHYSIOLOGY. Illustrated. Fcap. 8vo. 4s. 6d.
*QUESTIONS ON HUXLEY'S PHYSIOLOGY. By T. ALCOCK, M.D. 18mo. 1s. 6d.

HUXLEY and MARTIN.—A COURSE OF PRACTICAL INSTRUCTION IN ELEMENTARY BIOLOGY. By Prof. T. H. HUXLEY, F.R.S., assisted by H. N. MARTIN, F.R.S., Professor of Biology in the Johns Hopkins University, U.S.A. New Ed., revised and extended by G. B. HOWES and D. H. SCOTT, Ph.D., Assistant Professors, Normal School of Science and Royal School of Mines. With a Preface by T. H. HUXLEY, F.R.S. Cr. 8vo. 10s. 6d.

KLEIN.—Works by E. KLEIN, F.R.S., Lecturer on General Anatomy and Physiology in the Medical School of St. Bartholomew's Hospital, Professor of Bacteriology at the College of State Medicine, London.
MICRO-ORGANISMS AND DISEASE. An Introduction into the Study of Specific Micro-Organisms. Illustrated. 3d Ed., revised. Cr. 8vo. 6s.
THE BACTERIA IN ASIATIC CHOLERA. Cr. 8vo. 5s.

LANG.—TEXT-BOOK OF COMPARATIVE ANATOMY. By Dr. ARNOLD LANG. Professor of Zoology in the University of Zurich. Translated by H. M. BERNARD, M.A., and M. BERNARD. Introduction by Prof. E. HAECKEL. 2 vols. Illustrated. 8vo. Part I. 17s. net. [Part II. In the Press.

LANKESTER.—Works by E. RAY LANKESTER, F.R.S., Linacre Professor of Human and Comparative Anatomy in the University of Oxford.
A TEXT-BOOK OF ZOOLOGY. 8vo. [In preparation.
THE ADVANCEMENT OF SCIENCE. Occasional Essays and Addresses. 8vo. 10s. 6d.

LUBBOCK.—Works by the Right Hon. Sir JOHN LUBBOCK, F.R.S., D.C.L.
 THE ORIGIN AND METAMORPHOSES OF INSECTS. Illustrated. Cr. 8vo. 3s. 6d.
 ON BRITISH WILD FLOWERS CONSIDERED IN RELATION TO INSECTS. Illustrated. Cr. 8vo. 4s. 6d.
 FLOWERS, FRUITS, AND LEAVES. Illustrated. 2d Ed. Cr. 8vo. 4s. 6d.
 SCIENTIFIC LECTURES. 2d Ed. 8vo. 8s. 6d.
 FIFTY YEARS OF SCIENCE. Being the Address delivered at York to the British Association, August 1881. 5th Ed. Cr. 8vo. 2s. 6d.

MARTIN and MOALE.—ON THE DISSECTION OF VERTEBRATE ANIMALS. By Prof. H. N. MARTIN and W. A. MOALE. Cr. 8vo. [*In preparation.*

MIVART.—LESSONS IN ELEMENTARY ANATOMY. By ST. GEORGE MIVART, F.R.S., Lecturer on Comparative Anatomy at St. Mary's Hospital. Illustrated. Fcap. 8vo. 6s. 6d.

MÜLLER.—THE FERTILISATION OF FLOWERS. By HERMANN MÜLLER. Translated and Edited by D'ARCY W. THOMPSON, B.A., Professor of Biology in University College, Dundee. With a Preface by C. DARWIN, F.R.S. Illustrated. 8vo. 21s.

OLIVER.—Works by DANIEL OLIVER, F.R.S., late Professor of Botany in University College, London.
 *LESSONS IN ELEMENTARY BOTANY. Illustrated. Fcap. 8vo. 4s. 6d.
 FIRST BOOK OF INDIAN BOTANY. Illustrated. Ex. fcap. 8vo. 6s. 6d.

PARKER.—Works by T. JEFFERY PARKER, F.R.S., Professor of Biology in the University of Otago, New Zealand.
 A COURSE OF INSTRUCTION IN ZOOTOMY (VERTEBRATA). Illustrated. Cr. 8vo. 8s. 6d.
 LESSONS IN ELEMENTARY BIOLOGY. Illustrated. Cr. 8vo. 10s. 6d.

PARKER and BETTANY.—THE MORPHOLOGY OF THE SKULL. By Prof. W. K. PARKER, F.R.S., and G. T. BETTANY. Illustrated. Cr. 8vo. 10s. 6d.

ROMANES.—THE SCIENTIFIC EVIDENCES OF ORGANIC EVOLUTION. By GEORGE J. ROMANES, F.R.S., Zoological Secretary of the Linnean Society. Cr. 8vo. 2s. 6d.

SEDGWICK.—TREATISE ON EMBRYOLOGY. By ADAM SEDGWICK, F.R.S., Fellow and Lecturer of Trinity College, Cambridge. Illustrated. 8vo.
                  [*In preparation.*

SHUFELDT.—THE MYOLOGY OF THE RAVEN (*Corvus corax sinuatus*). A Guide to the Study of the Muscular System in Birds. By R. W. SHUFELDT. Illustrated. 8vo. 13s. net.

SMITH.—DISEASES OF FIELD AND GARDEN CROPS, CHIEFLY SUCH AS ARE CAUSED BY FUNGI. By W. G. SMITH, F.L.S. Illustrated. Fcap. 8vo. 4s. 6d.

STEWART and CORRY.—A FLORA OF THE NORTH-EAST OF IRELAND. Including the Phanerogamia, the Cryptogamia Vascularia, and the Muscineæ. By S. A. STEWART, Curator of the Collections in the Belfast Museum, and the late T. H. CORRY, M.A., Lecturer on Botany in the University Medical and Science Schools, Cambridge. Cr. 8vo. 5s. 6d.

WALLACE.—DARWINISM: An Exposition of the Theory of Natural Selection, with some of its Applications. By ALFRED RUSSEL WALLACE, LL.D., F.R.S. 3d Ed. Cr. 8vo. 9s.
 NATURAL SELECTION; AND TROPICAL NATURE. By the same. New Ed. Ex. Cr. 8vo. 6s.
 ISLAND LIFE. By the same. New Ed. Ex. Cr. 8vo. 6s.

WARD.—TIMBER AND SOME OF ITS DISEASES. By H. MARSHALL WARD, F.R.S., Professor of Botany in the Royal Indian Engineering College, Cooper's Hill. Illustrated. Cr. 8vo. 6s.

WIEDERSHEIM.—ELEMENTS OF THE COMPARATIVE ANATOMY OF VERTEBRATES. By Prof. R. WIEDERSHEIM. Adapted by W. NEWTON PARKER, Professor of Biology in the University College of South Wales and Monmouthshire. With Additions. Illustrated. 8vo. 12s. 6d.

MEDICINE.

BLYTH.—A MANUAL OF PUBLIC HEALTH. By A. WYNTER BLYTH, M.R.C.S. 8vo. 17s. net.

BRUNTON.—Works by T. LAUDER BRUNTON, M.D., F.R.S., Examiner in Materia Medica in the University of London, in the Victoria University, and in the Royal College of Physicians, London.
 A TEXT-BOOK OF PHARMACOLOGY, THERAPEUTICS, AND MATERIA MEDICA. Adapted to the United States Pharmacopœia by F. H. WILLIAMS, M.D., Boston, Mass. 3d Ed. Adapted to the New British Pharmacopœia, 1885, and additions, 1891. 8vo. 21s. Or in 2 Vols. 22s. 6d.
 TABLES OF MATERIA MEDICA: A Companion to the Materia Medica Museum. Illustrated. Cheaper Issue. 8vo. 5s.
 ON THE CONNECTION BETWEEN CHEMICAL CONSTITUTION AND PHYSIOLOGICAL ACTION, BEING AN INTRODUCTION TO MODERN THERAPEUTICS. Croonian Lectures. 8vo. [*In the Press.*

GRIFFITHS.—LESSONS ON PRESCRIPTIONS AND THE ART OF PRESCRIBING. By W. HANDSEL GRIFFITHS. Adapted to the Pharmacopœia, 1885. 18mo. 3s. 6d.

HAMILTON.—A TEXT-BOOK OF PATHOLOGY, SYSTEMATIC AND PRACTICAL. By D. J. HAMILTON, F.R.S.E., Professor of Pathological Anatomy, University of Aberdeen. Illustrated. Vol. I. 8vo. 25s.

KLEIN.—Works by E. KLEIN, F.R.S., Lecturer on General Anatomy and Physiology in the Medical School of St. Bartholomew's Hospital, London.
 MICRO-ORGANISMS AND DISEASE. An Introduction into the Study of Specific Micro-Organisms. Illustrated. 3d Ed., revised. Cr. 8vo. 6s.
 THE BACTERIA IN ASIATIC CHOLERA. Cr. 8vo. 5s.

WHITE.—A TEXT-BOOK OF GENERAL THERAPEUTICS. By W. HALE WHITE, M.D., Senior Assistant Physician to and Lecturer in Materia Medica at Guy's Hospital. Illustrated. Cr. 8vo. 8s. 6d.

ZIEGLER—MACALISTER.—TEXT-BOOK OF PATHOLOGICAL ANATOMY AND PATHOGENESIS. By Prof. E. ZIEGLER. Translated and Edited by DONALD MACALISTER, M.A., M.D., Fellow and Medical Lecturer of St. John's College, Cambridge. Illustrated. 8vo.
 Part I.—GENERAL PATHOLOGICAL ANATOMY. 2d Ed. 12s. 6d.
 Part II.—SPECIAL PATHOLOGICAL ANATOMY. Sections I.-VIII. 2d Ed. 12s. 6d. Sections IX.-XII. 12s. 6d.

HUMAN SCIENCES.

Mental and Moral Philosophy; Political Economy; Law and Politics; Anthropology; Education.

MENTAL AND MORAL PHILOSOPHY.

BALDWIN.—HANDBOOK OF PSYCHOLOGY: SENSES AND INTELLECT. By Prof. J. M. BALDWIN, M.A., LL.D. 2d Ed., revised. 8vo. 12s. 6d.

BOOLE.—THE MATHEMATICAL ANALYSIS OF LOGIC. Being an Essay towards a Calculus of Deductive Reasoning. By GEORGE BOOLE. 8vo. 5s.

CALDERWOOD.—HANDBOOK OF MORAL PHILOSOPHY. By Rev. HENRY CALDERWOOD, LL.D., Professor of Moral Philosophy in the University of Edinburgh. 14th Ed., largely rewritten. Cr. 8vo. 6s.

CLIFFORD.—SEEING AND THINKING. By the late Prof. W. K. CLIFFORD, F.R.S. With Diagrams. Cr. 8vo. 3s. 6d.

HÖFFDING.—OUTLINES OF PSYCHOLOGY. By Prof. H. HÖFFDING. Translated by M. E. LOWNDES. Cr. 8vo. 6s.

JAMES.—THE PRINCIPLES OF PSYCHOLOGY. By WM. JAMES, Professor of Psychology in Harvard University. 2 vols. 8vo. 25s. net.

JARDINE.—THE ELEMENTS OF THE PSYCHOLOGY OF COGNITION. By Rev. ROBERT JARDINE, D.Sc. 3d Ed., revised. Cr. 8vo. 6s. 6d.

JEVONS.—Works by W. STANLEY JEVONS, F.R.S.
*PRIMER OF LOGIC. 18mo. 1s.
*ELEMENTARY LESSONS IN LOGIC, Deductive and Inductive, with Copious Questions and Examples, and a Vocabulary of Logical Terms. Fcap. 8vo. 3s. 6d.
THE PRINCIPLES OF SCIENCE. A Treatise on Logic and Scientific Method. New and revised Ed. Cr. 8vo. 12s. 6d.
STUDIES IN DEDUCTIVE LOGIC. 2d Ed. Cr. 8vo. 6s.
PURE LOGIC: AND OTHER MINOR WORKS. Edited by R. ADAMSON, M.A., LL.D., Professor of Logic at Owens College, Manchester, and HARRIET A. JEVONS. With a Preface by Prof. ADAMSON. 8vo. 10s. 6d.

KANT—MAX MÜLLER.—CRITIQUE OF PURE REASON. By IMMANUEL KANT. 2 vols. 8vo. 16s. each. Vol. I. HISTORICAL INTRODUCTION, by LUDWIG NOIRÉ; Vol. II. CRITIQUE OF PURE REASON, translated by F. MAX MÜLLER.

KANT—MAHAFFY and BERNARD.—KANT'S CRITICAL PHILOSOPHY FOR ENGLISH READERS. By J. P. MAHAFFY, D.D., Professor of Ancient History in the University of Dublin, and JOHN H. BERNARD, B.D., Fellow of Trinity College, Dublin. A new and complete Edition in 2 vols. Cr. 8vo.
Vol. I. THE KRITIK OF PURE REASON EXPLAINED AND DEFENDED. 7s. 6d.
Vol. II. THE PROLEGOMENA. Translated with Notes and Appendices. 6s.

KEYNES.—FORMAL LOGIC, Studies and Exercises in. Including a Generalisation of Logical Processes in their application to Complex Inferences. By JOHN NEVILLE KEYNES, D.Sc. 2d Ed., revised and enlarged. Cr. 8vo. 10s. 6d.

McCOSH.—Works by JAMES McCOSH, D.D., President of Princeton College.
PSYCHOLOGY. Cr. 8vo.
I. THE COGNITIVE POWERS. 6s. 6d.
II. THE MOTIVE POWERS. 6s. 6d.
FIRST AND FUNDAMENTAL TRUTHS: being a Treatise on Metaphysics. Ex. cr. 8vo. 9s.
THE PREVAILING TYPES OF PHILOSOPHY. CAN THEY LOGICALLY REACH REALITY? 8vo. 3s. 6d.

MAURICE.—MORAL AND METAPHYSICAL PHILOSOPHY. By F. D. MAURICE, M.A., late Professor of Moral Philosophy in the University of Cambridge. Vol. I.—Ancient Philosophy and the First to the Thirteenth Centuries. Vol. II.—Fourteenth Century and the French Revolution, with a glimpse into the Nineteenth Century. 4th Ed. 2 vols. 8vo. 16s.

*RAY.—A TEXT-BOOK OF DEDUCTIVE LOGIC FOR THE USE OF STUDENTS. By P. K. RAY, D.Sc., Professor of Logic and Philosophy, Presidency College, Calcutta. 4th Ed. Globe 8vo. 4s. 6d.

SIDGWICK.—Works by HENRY SIDGWICK, LL.D., D.C.L., Knightbridge Professor of Moral Philosophy in the University of Cambridge.
THE METHODS OF ETHICS. 4th Ed. 8vo. 14s. A Supplement to the 2d Ed., containing all the important Additions and Alterations in the 3d Ed. 8vo. 6s.
OUTLINES OF THE HISTORY OF ETHICS, for English Readers. 2d Ed., revised. Cr. 8vo. 3s. 6d.

VENN.—Works by JOHN VENN, F.R.S., Examiner in Moral Philosophy in the University of London.
THE LOGIC OF CHANCE. An Essay on the Foundations and Province of the Theory of Probability, with special Reference to its Logical Bearings and its Application to Moral and Social Science. 3d Ed., rewritten and greatly enlarged. Cr. 8vo. 10s. 6d.
SYMBOLIC LOGIC. Cr. 8vo. 10s. 6d.
THE PRINCIPLES OF EMPIRICAL OR INDUCTIVE LOGIC. 8vo. 18s.

POLITICAL ECONOMY.

BASTABLE.—PUBLIC FINANCE. By C. F. BASTABLE, Professor of Political Economy at Trinity College, Dublin. 8vo. [In the Press.

BÖHM-BAWERK.—CAPITAL AND INTEREST. Translated by WILLIAM SMART, M.A. 8vo. 12s. net.

THE POSITIVE THEORY OF CAPITAL. By the same Author and Translator. 8vo. 12s. net.

CAIRNES.—THE CHARACTER AND LOGICAL METHOD OF POLITICAL ECONOMY. By J. E. CAIRNES. Cr. 8vo. 6s.

SOME LEADING PRINCIPLES OF POLITICAL ECONOMY NEWLY EXPOUNDED. By the same. 8vo. 14s.

COSSA.—GUIDE TO THE STUDY OF POLITICAL ECONOMY. By Dr. L. COSSA. Translated. With a Preface by W. S. JEVONS, F.R.S. Cr. 8vo. 4s. 6d.

*****FAWCETT.**—POLITICAL ECONOMY FOR BEGINNERS, WITH QUESTIONS. By Mrs. HENRY FAWCETT. 7th Ed. 18mo. 2s. 6d.

FAWCETT.—A MANUAL OF POLITICAL ECONOMY. By the Right Hon. HENRY FAWCETT, F.R.S. 7th Ed., revised. With a Chapter on "State Socialism and the Nationalisation of the Land," and an Index. Cr. 8vo. 12s. 6d.

AN EXPLANATORY DIGEST of the above. By C. A. WATERS, B.A. Cr. 8vo. 2s. 6d.

GILMAN.—PROFIT-SHARING BETWEEN EMPLOYER AND EMPLOYEE. A Study in the Evolution of the Wages System. By N. P. GILMAN. Cr. 8vo. 7s. 6d.

GUNTON.—WEALTH AND PROGRESS: A Critical Examination of the Wages Question and its Economic Relation to Social Reform. By GEORGE GUNTON. Cr. 8vo. 6s.

HOWELL.—THE CONFLICTS OF CAPITAL AND LABOUR HISTORICALLY AND ECONOMICALLY CONSIDERED. Being a History and Review of the Trade Unions of Great Britain, showing their Origin, Progress, Constitution, and Objects, in their varied Political, Social, Economical, and Industrial Aspects. By GEORGE HOWELL, M.P. 2d Ed., revised. Cr. 8vo. 7s. 6d.

JEVONS.—Works by W. STANLEY JEVONS, F.R.S.

*PRIMER OF POLITICAL ECONOMY. 18mo. 1s.

THE THEORY OF POLITICAL ECONOMY. 3d Ed., revised. 8vo. 10s. 6d.

KEYNES.—THE SCOPE AND METHOD OF POLITICAL ECONOMY. By J. N. KEYNES, D.Sc. 7s. net.

MARSHALL.—PRINCIPLES OF ECONOMICS. By ALFRED MARSHALL, M.A. 2 vols. 8vo. Vol. I. 2d Ed. 12s. 6d. net.

MARSHALL.—THE ECONOMICS OF INDUSTRY. By A. MARSHALL, M.A., Professor of Political Economy in the University of Cambridge, and MARY P. MARSHALL. Ex. fcap. 8vo. 2s. 6d.

PALGRAVE.—A DICTIONARY OF POLITICAL ECONOMY. By various Writers. Edited by R. H. INGLIS PALGRAVE, F.R.S. 3s. 6d. each, net. No. I. July 1891.

PANTALEONI.—MANUAL OF POLITICAL ECONOMY. By Prof. M. PANTALEONI. Translated by T. BOSTON BRUCE. [In preparation.

SIDGWICK.—THE PRINCIPLES OF POLITICAL ECONOMY. By HENRY SIDGWICK, LL.D., D.C.L., Knightbridge Professor of Moral Philosophy in the University of Cambridge. 2d Ed., revised. 8vo. 16s.

SMART.—AN INTRODUCTION TO THE THEORY OF VALUE. By WILLIAM SMART, M.A. Crown 8vo. [In the Press.

WALKER.—Works by FRANCIS A. WALKER, M.A.
FIRST LESSONS IN POLITICAL ECONOMY. Cr. 8vo. 5s.
A BRIEF TEXT-BOOK OF POLITICAL ECONOMY. Cr. 8vo. 6s. 6d.
POLITICAL ECONOMY. 2d Ed., revised and enlarged. 8vo. 12s. 6d.
THE WAGES QUESTION. Ex. Cr. 8vo. 8s. 6d. net.
MONEY. Ex. Cr. 8vo. 8s. 6d. net.

*****WICKSTEED.**—ALPHABET OF ECONOMIC SCIENCE. By PHILIP H. WICKSTEED, M.A. Part I. Elements of the Theory of Value or Worth. Gl. 8vo. 2s. 6d.

HUMAN SCIENCES
LAW AND POLITICS.

ADAMS and CUNNINGHAM.—THE SWISS CONFEDERATION. By Sir F. O. ADAMS and C. CUNNINGHAM. 8vo. 14s.

ANGLO-SAXON LAW, ESSAYS ON.—Contents: Anglo-Saxon Law Courts, Land and Family Law, and Legal Procedure. 8vo. 18s.

BALL.—THE STUDENT'S GUIDE TO THE BAR. By WALTER W. R. BALL, M.A., Fellow and Assistant Tutor of Trinity College, Cambridge. 4th Ed., revised. Cr. 8vo. 2s. 6d.

BIGELOW.—HISTORY OF PROCEDURE IN ENGLAND FROM THE NORMAN CONQUEST. The Norman Period, 1066-1204. By MELVILLE M. BIGELOW, Ph.D., Harvard University. 8vo. 16s.

BOUTMY. — STUDIES IN CONSTITUTIONAL LAW. By EMILE BOUTMY. Translated by Mrs. DICEY, with Preface by Prof. A. V. DICEY. Cr. 8vo. 6s.

THE ENGLISH CONSTITUTION. By the same. Translated by Mrs. EADEN, with Introduction by Sir F. POLLOCK, Bart. Cr. 8vo. 6s.

BRYCE.—THE AMERICAN COMMONWEALTH. By JAMES BRYCE, M.P., D.C.L., Regius Professor of Civil Law in the University of Oxford. Two Volumes. Ex. cr. 8vo. 25s. Part I. The National Government. Part II. The State Governments. Part III. The Party System. Part IV. Public Opinion. Part V. Illustrations and Reflections. Part VI. Social Institutions.

*****BUCKLAND.**—OUR NATIONAL INSTITUTIONS. A Short Sketch for Schools. By ANNA BUCKLAND. With Glossary. 18mo. 1s.

CHERRY.—LECTURES ON THE GROWTH OF CRIMINAL LAW IN ANCIENT COMMUNITIES. By R. R. CHERRY, LL.D., Reid Professor of Constitutional and Criminal Law in the University of Dublin. 8vo. 5s. net.

DICEY.—INTRODUCTION TO THE STUDY OF THE LAW OF THE CONSTITUTION. By A. V. DICEY, B.C.L., Vinerian Professor of English Law in the University of Oxford. 3d Ed. 8vo. 12s. 6d.

DILKE.—PROBLEMS OF GREATER BRITAIN. By the Right Hon. Sir CHARLES WENTWORTH DILKE. With Maps. 4th Ed. Ex. cr. 8vo. 12s. 6d.

DONISTHORPE.—INDIVIDUALISM: A System of Politics. By WORDSWORTH DONISTHORPE. 8vo. 14s.

ENGLISH CITIZEN, THE.—A Series of Short Books on his Rights and Responsibilities. Edited by HENRY CRAIK, LL.D. Cr. 8vo. 3s. 6d. each.
CENTRAL GOVERNMENT. By H. D. TRAILL, D.C.L.
THE ELECTORATE AND THE LEGISLATURE. By SPENCER WALPOLE.
THE POOR LAW. By Rev. T. W. FOWLE, M.A. New Ed. With Appendix.
THE NATIONAL BUDGET; THE NATIONAL DEBT; TAXES AND RATES. By A. J. WILSON.
THE STATE IN RELATION TO LABOUR. By W. STANLEY JEVONS, LL.D.
THE STATE AND THE CHURCH. By the Hon. ARTHUR ELLIOT.
FOREIGN RELATIONS. By SPENCER WALPOLE.
THE STATE IN ITS RELATION TO TRADE. By Sir T. H. FARRER, Bart.
LOCAL GOVERNMENT. By M. D. CHALMERS, M.A.
THE STATE IN ITS RELATION TO EDUCATION. By HENRY CRAIK, LL.D.
THE LAND LAWS. By Sir F. POLLOCK, Bart., Professor of Jurisprudence in the University of Oxford.
COLONIES AND DEPENDENCIES. Part I. INDIA. By J. S. COTTON, M.A. II. THE COLONIES. By E. J. PAYNE, M.A.
JUSTICE AND POLICE. By F. W. MAITLAND.
THE PUNISHMENT AND PREVENTION OF CRIME. By Colonel Sir EDMUND DU CANE, K.C.B., Chairman of Commissioners of Prisons.

FISKE.—CIVIL GOVERNMENT IN THE UNITED STATES CONSIDERED WITH SOME REFERENCE TO ITS ORIGINS. By JOHN FISKE, formerly Lecturer on Philosophy at Harvard University. Cr. 8vo. 6s. 6d.

HOLMES.—THE COMMON LAW. By O. W. HOLMES, Jun. Demy 8vo. 12s.

JENKS.—THE GOVERNMENT OF VICTORIA. By EDWARD JENKS, B.A., LL.B., Professor of Law in the University of Melbourne. [*In the Press.*

MAITLAND.—PLEAS OF THE CROWN FOR THE COUNTY OF GLOUCESTER

BEFORE THE ABBOT OF READING AND HIS FELLOW JUSTICES ITINERANT, IN THE FIFTH YEAR OF THE REIGN OF KING HENRY THE THIRD, AND THE YEAR OF GRACE 1221. By F. W. MAITLAND. 8vo. 7s. 6d.

MUNRO.—COMMERCIAL LAW. By J. E. C. MUNRO, LL.D., Professor of Law and Political Economy in the Owens College, Manchester. [In preparation.

PATERSON.—Works by JAMES PATERSON, Barrister-at-Law.
COMMENTARIES ON THE LIBERTY OF THE SUBJECT, AND THE LAWS OF ENGLAND RELATING TO THE SECURITY OF THE PERSON. Cheaper Issue. Two Vols. Cr. 8vo. 21s.
THE LIBERTY OF THE PRESS, SPEECH, AND PUBLIC WORSHIP. Being Commentaries on the Liberty of the Subject and the Laws of England. Cr. 8vo. 12s.

PHILLIMORE.—PRIVATE LAW AMONG THE ROMANS. From the Pandects. By J. G. PHILLIMORE, Q.C. 8vo. 16s.

POLLOCK.—ESSAYS IN JURISPRUDENCE AND ETHICS. By Sir FREDERICK POLLOCK, Bart., Corpus Christi Professor of Jurisprudence in the University of Oxford. 8vo. 10s. 6d.
INTRODUCTION TO THE HISTORY OF THE SCIENCE OF POLITICS. By the same. Cr. 8vo. 2s. 6d.

RICHEY.—THE IRISH LAND LAWS. By ALEX. G. RICHEY, Q.C., Deputy Regius Professor of Feudal English Law in the University of Dublin. Cr. 8vo. 3s. 6d.

SIDGWICK.—THE ELEMENTS OF POLITICS. By HENRY SIDGWICK, LL.D. 8vo. 14s. net.

STEPHEN.—Works by Sir J. FITZJAMES STEPHEN, Bart.
A DIGEST OF THE LAW OF EVIDENCE. 5th Ed., revised and enlarged. Cr. 8vo. 6s.
A DIGEST OF THE CRIMINAL LAW: CRIMES AND PUNISHMENTS. 4th Ed., revised. 8vo. 16s.
A DIGEST OF THE LAW OF CRIMINAL PROCEDURE IN INDICTABLE OFFENCES. By Sir J. F. STEPHEN, Bart., and H. STEPHEN, LL.M., of the Inner Temple, Barrister-at-Law. 8vo. 12s. 6d.
A HISTORY OF THE CRIMINAL LAW OF ENGLAND. Three Vols. 8vo. 48s.
GENERAL VIEW OF THE CRIMINAL LAW OF ENGLAND. 8vo. 14s.

ANTHROPOLOGY.

DAWKINS.—EARLY MAN IN BRITAIN AND HIS PLACE IN THE TERTIARY PERIOD. By Prof. W. BOYD DAWKINS. Medium 8vo. 25s.

FRAZER.—THE GOLDEN BOUGH. A Study in Comparative Religion. By J. G. FRAZER, M.A., Fellow of Trinity College, Cambridge. 2 vols. 8vo. 28s.

M'LENNAN.—THE PATRIARCHAL THEORY. Based on the papers of the late JOHN F. M'LENNAN. Edited by DONALD M'LENNAN, M.A., Barrister-at-Law. 8vo. 14s.
STUDIES IN ANCIENT HISTORY. By the same. Comprising a Reprint of "Primitive Marriage." An Inquiry into the origin of the form of capture in Marriage Ceremonies. 8vo. 16s.

TYLOR.—ANTHROPOLOGY. An Introduction to the Study of Man and Civilisation. By E. B. TYLOR, F.R.S. Illustrated. Cr. 8vo. 7s. 6d.

WESTERMARCK.—THE HISTORY OF HUMAN MARRIAGE. By Dr. EDWARD WESTERMARCK. With Preface by A. R. WALLACE. 8vo. 14s. net.

WILSON.—THE RIGHT HAND: LEFT-HANDEDNESS. By Sir D. WILSON. Cr. 8vo. 4s. 6d.

EDUCATION.

ARNOLD.—REPORTS ON ELEMENTARY SCHOOLS. 1852-1882. By MATTHEW ARNOLD, D.C.L. Edited by the Right Hon. Sir FRANCIS SANDFORD, K.C.B. Cheaper Issue. Cr. 8vo. 3s. 6d.

HIGHER SCHOOLS AND UNIVERSITIES IN GERMANY. By the same. Crown 8vo. 6s.

BALL.—THE STUDENT'S GUIDE TO THE BAR. By WALTER W. R. BALL, M.A., Fellow and Assistant Tutor of Trinity College, Cambridge. 4th Ed., revised. Cr. 8vo. 2s. 6d.

*BLAKISTON.—THE TEACHER. Hints on School Management. A Handbook for Managers, Teachers' Assistants, and Pupil Teachers. By J. R. BLAKISTON. Cr. 8vo. 2s. 6d. (Recommended by the London, Birmingham, and Leicester School Boards.)

CALDERWOOD.—ON TEACHING. By Prof. HENRY CALDERWOOD. New Ed. Ex. fcap. 8vo. 2s. 6d.

FEARON.—SCHOOL INSPECTION. By D. R. FEARON. 6th Ed. Cr. 8vo. 2s. 6d.

FITCH.—NOTES ON AMERICAN SCHOOLS AND TRAINING COLLEGES. Reprinted from the Report of the English Education Department for 1888-89, with permission of the Controller of H.M.'s Stationery Office. By J. G. FITCH, M.A. Gl. 8vo. 2s. 6d.

GEIKIE.—THE TEACHING OF GEOGRAPHY. A Practical Handbook for the use of Teachers. By Sir ARCHIBALD GEIKIE, F.R.S., Director-General of the Geological Survey of the United Kingdom. Cr. 8vo. 2s.

GLADSTONE.—SPELLING REFORM FROM A NATIONAL POINT OF VIEW. By J. H. GLADSTONE. Cr. 8vo. 1s. 6d.

HERTEL.—OVERPRESSURE IN HIGH SCHOOLS IN DENMARK. By Dr. HERTEL. Translated by C. G. SÖRENSEN. With Introduction by Sir J. CRICHTON-BROWNE, F.R.S. Cr. 8vo. 3s. 6d.

TODHUNTER.—THE CONFLICT OF STUDIES. By ISAAC TODHUNTER, F.R.S. 8vo. 10s. 6d.

TECHNICAL KNOWLEDGE.

(See also MECHANICS, LAW, and MEDICINE.)

Civil and Mechanical Engineering; Military and Naval Science; Agriculture; Domestic Economy; Book-Keeping; Commerce.

CIVIL AND MECHANICAL ENGINEERING.

ALEXANDER and THOMSON.—ELEMENTARY APPLIED MECHANICS. By T. ALEXANDER, Professor of Civil Engineering, Trinity College, Dublin, and A. W. THOMSON, Professor at College of Science, Poona, India. Part II. TRANSVERSE STRESS. Cr. 8vo. 10s. 6d.

CHALMERS.—GRAPHICAL DETERMINATION OF FORCES IN ENGINEERING STRUCTURES. By J. B. CHALMERS, C.E. Illustrated. 8vo. 24s.

COTTERILL.—APPLIED MECHANICS: An Elementary General Introduction to the Theory of Structures and Machines. By J. H. COTTERILL, F.R.S., Professor of Applied Mechanics in the Royal Naval College, Greenwich. 2d Ed. 8vo. 18s.

COTTERILL and SLADE.—LESSONS IN APPLIED MECHANICS. By Prof. J. H. COTTERILL and J. H. SLADE. Fcap. 8vo. 5s. 6d.

GRAHAM.—GEOMETRY OF POSITION. By R. H. GRAHAM. Cr. 8vo. 7s. 6d.

KENNEDY.—THE MECHANICS OF MACHINERY. By A. B. W. KENNEDY, F.R.S. Illustrated. Cr. 8vo. 12s. 6d.

WHITHAM.—STEAM-ENGINE DESIGN. For the Use of Mechanical Engineers, Students, and Draughtsmen. By J. M. WHITHAM, Professor of Engineering, Arkansas Industrial University. Illustrated. 8vo. 25s.

YOUNG.—SIMPLE PRACTICAL METHODS OF CALCULATING STRAINS ON GIRDERS, ARCHES, AND TRUSSES. With a Supplementary Essay on Economy in Suspension Bridges. By E. W. YOUNG, C.E. With Diagrams. 8vo. 7s. 6d.

MILITARY AND NAVAL SCIENCE.

AITKEN.—THE GROWTH OF THE RECRUIT AND YOUNG SOLDIER. With a view to the selection of "Growing Lads" for the Army, and a Regulated

MILITARY SCIENCE—AGRICULTURE

System of Training for Recruits. By Sir W. AITKEN, F.R.S., Professor of Pathology in the Army Medical School. Cr. 8vo. 8s. 6d.

ARMY PRELIMINARY EXAMINATION, 1882-1890, Specimens of Papers set at the. With Answers to the Mathematical Questions. Subjects: Arithmetic, Algebra, Euclid, Geometrical Drawing, Geography, French, English Dictation. Cr. 8vo. 3s. 6d.

MATTHEWS.—MANUAL OF LOGARITHMS. By G. F. MATTHEWS, B.A. 8vo. 5s. net.

MAURICE.—WAR. By FREDERICK MAURICE, Colonel C.B., R.A. 8vo. 5s. net.

MERCUR.—ELEMENTS OF THE ART OF WAR. Prepared for the use of Cadets of the United States Military Academy. By JAMES MERCUR, Professor of Civil Engineering at the United States Academy, West Point, New York. 2d Ed., revised and corrected. 8vo. 17s.

PALMER.—TEXT-BOOK OF PRACTICAL LOGARITHMS AND TRIGONOMETRY. By J. H. PALMER, Head Schoolmaster, R.N., H.M.S. *Cambridge*, Devonport. Gl. 8vo. 4s. 6d.

ROBINSON.—TREATISE ON MARINE SURVEYING. Prepared for the use of younger Naval Officers. With Questions for Examinations and Exercises principally from the Papers of the Royal Naval College. With the results. By Rev. JOHN L. ROBINSON, Chaplain and Instructor in the Royal Naval College, Greenwich. Illustrated. Cr. 8vo. 7s. 6d.

SANDHURST MATHEMATICAL PAPERS, for Admission into the Royal Military College, 1881-1889. Edited by E. J. BROOKSMITH, B.A., Instructor in Mathematics at the Royal Military Academy, Woolwich. Cr. 8vo. 3s. 6d.

SHORTLAND.—NAUTICAL SURVEYING. By the late Vice-Admiral SHORTLAND, LL.D. 8vo. 21s.

THOMSON.—POPULAR LECTURES AND ADDRESSES. By Sir WILLIAM THOMSON, LL.D., P.R.S. In 3 vols. Illustrated. Cr. 8vo. Vol. III. Navigation. 7s. 6d.

WILKINSON.—THE BRAIN OF AN ARMY. A Popular Account of the German General Staff. By SPENSER WILKINSON. Cr. 8vo. 2s. 6d.

WOLSELEY.—Works by General Viscount WOLSELEY, G.C.M.G.
THE SOLDIER'S POCKET-BOOK FOR FIELD SERVICE. 5th Ed., revised and enlarged. 16mo. Roan. 5s.
FIELD POCKET-BOOK FOR THE AUXILIARY FORCES. 16mo. 1s. 6d.

WOOLWICH MATHEMATICAL PAPERS, for Admission into the Royal Military Academy, Woolwich, 1880-1888 inclusive. Edited by E. J. BROOKSMITH, B.A., Instructor in Mathematics at the Royal Military Academy, Woolwich. Cr. 8vo. 6s.

AGRICULTURE.

FRANKLAND.—AGRICULTURAL CHEMICAL ANALYSIS, A Handbook of. By PERCY F. FRANKLAND, F.R.S., Professor of Chemistry, University College, Dundee. Founded upon *Leitfaden für die Agriculture Chemiche Analyse*, von Dr. F. KROCKER. Cr. 8vo. 7s. 6d.

HARTIG.—TEXT-BOOK OF THE DISEASES OF TREES. By Dr. ROBERT HARTIG. Translated by WM. SOMERVILLE, B.Sc., D.Œ., Professor of Agriculture and Forestry, Durham College of Science, Newcastle-on-Tyne. Edited, with Introduction, by Prof. H. MARSHALL WARD. 8vo. [*In preparation.*

LASLETT.—TIMBER AND TIMBER TREES, NATIVE AND FOREIGN. By THOMAS LASLETT. Cr. 8vo. 8s. 6d.

SMITH.—DISEASES OF FIELD AND GARDEN CROPS, CHIEFLY SUCH AS ARE CAUSED BY FUNGI. By WORTHINGTON G. SMITH, F.L.S. Illustrated. Fcap. 8vo. 4s. 6d.

TANNER.—*ELEMENTARY LESSONS IN THE SCIENCE OF AGRICULTURAL PRACTICE. By HENRY TANNER, F.C.S., M.R.A.C., Examiner in the Principles of Agriculture under the Government Department of Science. Fcap. 8vo. 3s. 6d.
*FIRST PRINCIPLES OF AGRICULTURE. By the same. 18mo. 1s.
THE PRINCIPLES OF AGRICULTURE. By the same. A Series of Reading Books for use in Elementary Schools. Ex. fcap. 8vo.
*I. The Alphabet of the Principles of Agriculture. 6d.

*II. Further Steps in the Principles of Agriculture. 1s.
*III. Elementary School Readings on the Principles of Agriculture for the third stage. 1s.

WARD.—TIMBER AND SOME OF ITS DISEASES. By H. MARSHALL WARD, M.A., F.L.S., F.R.S., Fellow of Christ's College, Cambridge, Professor of Botany at the Royal Indian Engineering College, Cooper's Hill. With Illustrations. Cr. 8vo. 6s.

DOMESTIC ECONOMY.

*BARKER.—FIRST LESSONS IN THE PRINCIPLES OF COOKING. By LADY BARKER. 18mo. 1s.

*BERNERS.—FIRST LESSONS ON HEALTH. By J. BERNERS. 18mo. 1s.

*COOKERY BOOK.—THE MIDDLE CLASS COOKERY BOOK. Edited by the Manchester School of Domestic Cookery. Fcap. 8vo. 1s. 6d.

CRAVEN.—A GUIDE TO DISTRICT NURSES. By Mrs. DACRE CRAVEN (née FLORENCE SARAH LEES), Hon. Associate of the Order of St. John of Jerusalem, etc. Cr. 8vo. 2s. 6d.

FREDERICK.—HINTS TO HOUSEWIVES ON SEVERAL POINTS, PARTICULARLY ON THE PREPARATION OF ECONOMICAL AND TASTEFUL DISHES. By Mrs. FREDERICK. Cr. 8vo. 1s.

*GRAND'HOMME.—CUTTING-OUT AND DRESSMAKING. From the French of Mdlle. E. GRAND'HOMME. With Diagrams. 18mo. 1s.

JEX-BLAKE.—THE CARE OF INFANTS. A Manual for Mothers and Nurses. By SOPHIA JEX-BLAKE, M.D., Lecturer on Hygiene at the London School of Medicine for Women. 18mo. 1s.

RATHBONE.—THE HISTORY AND PROGRESS OF DISTRICT NURSING FROM ITS COMMENCEMENT IN THE YEAR 1859 TO THE PRESENT DATE, including the foundation by the Queen of the Queen Victoria Jubilee Institute for Nursing the Poor in their own Homes. By WILLIAM RATHBONE, M.P. Cr. 8vo. 2s. 6d.

*TEGETMEIER.—HOUSEHOLD MANAGEMENT AND COOKERY. With an Appendix of Recipes used by the Teachers of the National School of Cookery. By W. B. TEGETMEIER. Compiled at the request of the School Board for London. 18mo. 1s.

*WRIGHT.—THE SCHOOL COOKERY-BOOK. Compiled and Edited by C. E. GUTHRIE WRIGHT, Hon. Sec. to the Edinburgh School of Cookery. 18mo. 1s.

BOOK-KEEPING.

*THORNTON.—FIRST LESSONS IN BOOK-KEEPING. By J. THORNTON. Cr. 8vo. 2s. 6d. KEY. Oblong 4to. 10s. 6d.
*PRIMER OF BOOK-KEEPING. By the same. 18mo. 1s.
KEY. 8vo. 2s. 6d.

COMMERCE.

MACMILLAN'S ELEMENTARY COMMERCIAL CLASS BOOKS. Edited by JAMES GOW, Litt.D., Headmaster of Nottingham School. Globe 8vo.

The following volumes are arranged for :—

*THE HISTORY OF COMMERCE IN EUROPE. By H. DE B. GIBBINS, M.A. 3s. 6d. [*Ready.*
COMMERCIAL GERMAN. By F. C. SMITH, B.A., formerly scholar of Magdalene College, Cambridge. [*In the Press.*
COMMERCIAL GEOGRAPHY. By E. C. K. GONNER, M.A., Professor of Political Economy in University College, Liverpool. [*In preparation.*
COMMERCIAL FRENCH.

GEOGRAPHY—HISTORY

COMMERCIAL ARITHMETIC. By A. W. SUNDERLAND, M.A., late Scholar of Trinity College, Cambridge; Fellow of the Institute of Actuaries. [*In prep.*

COMMERCIAL LAW. By J. E. C. MUNRO, LL.D., Professor of Law and Political Economy in the Owens College, Manchester.

GEOGRAPHY.
(See also PHYSICAL GEOGRAPHY.)

BARTHOLOMEW.—*THE ELEMENTARY SCHOOL ATLAS. By JOHN BARTHOLOMEW, F.R.G.S. 4to. 1s.

*MACMILLAN'S SCHOOL ATLAS, PHYSICAL AND POLITICAL. Consisting of 80 Maps and complete Index. By the same. Prepared for the use of Senior Pupils. Royal 4to. 8s. 6d. Half-morocco. 10s. 6d.

THE LIBRARY REFERENCE ATLAS OF THE WORLD. By the same. A Complete Series of 84 Modern Maps. With Geographical Index to 100,000 places. Half-morocco. Gilt edges. Folio. £2:12:6 net. Also issued in parts, 5s. each net. Geographical Index, 7s. 6d. net. Part I., April 1891.

*CLARKE.—CLASS-BOOK OF GEOGRAPHY. By C. B. CLARKE, F.R.S. New Ed., revised 1889, with 18 Maps. Fcap. 8vo. 3s. Sewed, 2s. 6d.

*GREEN.—A SHORT GEOGRAPHY OF THE BRITISH ISLANDS. By JOHN RICHARD GREEN and A. S. GREEN. With Maps. Fcap. 8vo. 3s. 6d.

*GROVE.—A PRIMER OF GEOGRAPHY. By Sir GEORGE GROVE, D.C.L. Illustrated. 18mo. 1s.

KIEPERT.—A MANUAL OF ANCIENT GEOGRAPHY. By Dr. H. KIEPERT. Cr. 8vo. 5s.

MACMILLAN'S GEOGRAPHICAL SERIES.—Edited by Sir ARCHIBALD GEIKIE, F.R.S., Director-General of the Geological Survey of the United Kingdom.

*THE TEACHING OF GEOGRAPHY. A Practical Handbook for the Use of Teachers. By Sir ARCHIBALD GEIKIE, F.R.S. Cr. 8vo. 2s.

*MAPS AND MAP-DRAWING. By W. A. ELDERTON. 18mo. 1s.

*GEOGRAPHY OF THE BRITISH ISLES. By Sir A. GEIKIE, F.R.S. 18mo. 1s.

*AN ELEMENTARY CLASS-BOOK OF GENERAL GEOGRAPHY. By H. R. MILL, D.Sc., Lecturer on Physiography and on Commercial Geography in the Heriot-Watt College, Edinburgh. Illustrated. Cr. 8vo. 3s. 6d.

*GEOGRAPHY OF EUROPE. By J. SIME, M.A. Illustrated. Gl. 8vo. 3s.

*ELEMENTARY GEOGRAPHY OF INDIA, BURMA, AND CEYLON. By H. F. BLANFORD, F.G.S. Gl. 8vo. 2s. 6d.

GEOGRAPHY OF NORTH AMERICA. By Prof. N. S. SHALER. [*In preparation.*

GEOGRAPHY OF THE BRITISH COLONIES. By G. M. DAWSON and A. SUTHERLAND. [*In the Press.*

STRACHEY.—LECTURES ON GEOGRAPHY. By General RICHARD STRACHEY, R.E. Cr. 8vo. 4s. 6d.

*TOZER.—A PRIMER OF CLASSICAL GEOGRAPHY. By H. F. TOZER, M.A. 18mo. 1s.

HISTORY.

ARNOLD.—THE SECOND PUNIC WAR. Being Chapters from THE HISTORY OF ROME, by the late THOMAS ARNOLD, D.D., Headmaster of Rugby. Edited, with Notes, by W. T. ARNOLD, M.A. With 8 Maps. Cr. 8vo. 5s.

ARNOLD.—A HISTORY OF THE EARLY ROMAN EMPIRE. By W. T. ARNOLD, M.A. Cr. 8vo. [*In preparation.*

*BEESLY.—STORIES FROM THE HISTORY OF ROME. By Mrs. BEESLY. Fcap. 8vo. 2s. 6d.

BRYCE.—Works by JAMES BRYCE, M.P., D.C.L., Regius Professor of Civil Law in the University of Oxford.

THE HOLY ROMAN EMPIRE. 9th Ed. Cr. 8vo. 7s. 6d.
⁎ Also a *Library Edition*. Demy 8vo. 14s.

HISTORY

THE AMERICAN COMMONWEALTH. 2 vols. Ex. cr. 8vo. 25s. Part I. The National Government. Part II. The State Governments. Part III. The Party System. Part IV. Public Opinion. Part V. Illustrations and Reflections. Part VI. Social Institutions.

*BUCKLEY.—A HISTORY OF ENGLAND FOR BEGINNERS. By ARABELLA B. BUCKLEY. With Maps and Tables. Gl. 8vo. 3s.

BURY.—A HISTORY OF THE LATER ROMAN EMPIRE FROM ARCADIUS TO IRENE, A.D. 395-800. By JOHN B. BURY, M.A., Fellow of Trinity College, Dublin. 2 vols. 8vo. 32s.

CASSEL.—MANUAL OF JEWISH HISTORY AND LITERATURE. By Dr. D. CASSEL. Translated by Mrs. HENRY LUCAS. Fcap. 8vo. 2s. 6d.

ENGLISH STATESMEN, TWELVE. Cr. 8vo. 2s. 6d. each.
WILLIAM THE CONQUEROR. By EDWARD A. FREEMAN, D.C.L., LL.D.
HENRY II. By Mrs. J. R. GREEN.
EDWARD I. By F. YORK POWELL. [In preparation.
HENRY VII. By JAMES GAIRDNER.
CARDINAL WOLSEY. By Bishop CREIGHTON.
ELIZABETH. By E. S. BEESLY. [Nearly Ready.
OLIVER CROMWELL. By FREDERIC HARRISON.
WILLIAM III. By H. D. TRAILL.
WALPOLE. By JOHN MORLEY.
CHATHAM. By JOHN MORLEY. - [Nearly Ready.
PITT. By EARL OF ROSEBERY. [Nearly Ready.
PEEL. By J. R. THURSFIELD.

FISKE.—Works by JOHN FISKE, formerly Lecturer on Philosophy at Harvard University.
THE CRITICAL PERIOD IN AMERICAN HISTORY, 1783-1789. Ex. cr. 8vo. 10s. 6d.
THE BEGINNINGS OF NEW ENGLAND; or, The Puritan Theocracy in its Relations to Civil and Religious Liberty. Cr. 8vo. 7s. 6d.
THE AMERICAN REVOLUTION. 2 vols. Cr. 8vo. 18s.

FREEMAN.—Works by EDWARD A. FREEMAN, D.C.L., Regius Professor of Modern History in the University of Oxford, etc.
*OLD ENGLISH HISTORY. With Maps. Ex. fcap. 8vo. 6s.
A SCHOOL HISTORY OF ROME. Cr. 8vo. [In preparation.
METHODS OF HISTORICAL STUDY. 8vo. 10s. 6d.
THE CHIEF PERIODS OF EUROPEAN HISTORY. Six Lectures. With an Essay on Greek Cities under Roman Rule. 8vo. 10s. 6d.
HISTORICAL ESSAYS. First Series. 4th Ed. 8vo. 10s. 6d.
HISTORICAL ESSAYS. Second Series. 3d Ed., with additional Essays. 8vo. 10s. 6d.
HISTORICAL ESSAYS. Third Series. 8vo. 12s.
THE GROWTH OF THE ENGLISH CONSTITUTION FROM THE EARLIEST TIMES. 4th Ed. Cr. 8vo. 5s.
*GENERAL SKETCH OF EUROPEAN HISTORY. Enlarged, with Maps, etc. 18mo. 3s. 6d.
*PRIMER OF EUROPEAN HISTORY. 18mo. 1s. (*History Primers.*)

FRIEDMANN.—ANNE BOLEYN. A Chapter of English History, 1527-1536. By PAUL FRIEDMANN. 2 vols. 8vo. 28s.

*GIBBINS.—THE HISTORY OF COMMERCE IN EUROPE. By H. de B. GIBBINS, M.A. With Maps. Globe 8vo. 3s. 6d.

GREEN.—Works by JOHN RICHARD GREEN, LL.D., late Honorary Fellow of Jesus College, Oxford.
*A SHORT HISTORY OF THE ENGLISH PEOPLE. New and Revised Ed. With Maps, Genealogical Tables, and Chronological Annals. Cr. 8vo. 8s. 6d. 159th Thousand.
*Also the same in Four Parts. With the corresponding portion of Mr. Tait's "Analysis." Crown 8vo. 3s. each. Part I. 607-1265. Part II. 1204-1553. Part III. 1540-1689. Part IV. 1660-1873.

HISTORY

HISTORY OF THE ENGLISH PEOPLE. In four vols. 8vo. 16s. each.
  Vol. I.—Early England, 449-1071; Foreign Kings, 1071-1214; The Charter, 1214-1291; The Parliament, 1307-1461. With 8 Maps.
  Vol. II.—The Monarchy, 1461-1540; The Reformation, 1540-1603.
  Vol. III.—Puritan England, 1603-1660; The Revolution, 1660-1688. With four Maps.
  Vol. IV.—The Revolution, 1688-1760; Modern England, 1760-1815. With Maps and Index.
THE MAKING OF ENGLAND. With Maps. 8vo. 16s.
THE CONQUEST OF ENGLAND. With Maps and Portrait. 8vo. 18s.
*ANALYSIS OF ENGLISH HISTORY, based on Green's "Short History of the English People." By C. W. A. TAIT, M.A., Assistant Master at Clifton College. Revised and Enlarged Ed. Crown 8vo. 4s. 6d.
*READINGS FROM ENGLISH HISTORY. Selected and Edited by JOHN RICHARD GREEN. Three Parts. Gl. 8vo. 1s. 6d. each. I. Hengist to Cressy. II. Cressy to Cromwell. III. Cromwell to Balaklava.
GUEST.—LECTURES ON THE HISTORY OF ENGLAND. By M. J. GUEST. With Maps. Cr. 8vo. 6s.
*HISTORICAL COURSE FOR SCHOOLS.—Edited by E. A. FREEMAN, D.C.L., Regius Professor of Modern History in the University of Oxford. 18mo.
  GENERAL SKETCH OF EUROPEAN HISTORY. By E. A. FREEMAN, D.C.L. New Ed., revised and enlarged. With Chronological Table, Maps, and Index. 3s. 6d.
  HISTORY OF ENGLAND. By EDITH THOMPSON. New Ed., revised and enlarged. With Coloured Maps. 2s. 6d.
  HISTORY OF SCOTLAND. By MARGARET MACARTHUR. 2s.
  HISTORY OF ITALY. By Rev. W. HUNT, M.A. New Ed. With Coloured Maps. 3s. 6d.
  HISTORY OF GERMANY. By J. SIME, M.A. New Ed., revised. 3s.
  HISTORY OF AMERICA. By JOHN A. DOYLE. With Maps. 4s. 6d.
  HISTORY OF EUROPEAN COLONIES. By E. J. PAYNE, M.A. With Maps. 4s. 6d.
  HISTORY OF FRANCE. By CHARLOTTE M. YONGE. With Maps. 3s. 6d.
  HISTORY OF GREECE. By EDWARD A. FREEMAN, D.C.L. [In preparation.
  HISTORY OF ROME. By EDWARD A. FREEMAN, D.C.L. [In preparation.
*HISTORY PRIMERS.—Edited by JOHN RICHARD GREEN, LL.D. 18mo. 1s. each.
  ROME. By Bishop CREIGHTON. Maps.
  GREECE. By C. A. FYFFE, M.A., late Fellow of University College, Oxford. Maps.
  EUROPE. By E. A. FREEMAN, D.C.L. Maps.
  FRANCE. By CHARLOTTE M. YONGE.
  GREEK ANTIQUITIES. By Rev. J. P. MAHAFFY, D.D. Illustrated.
  CLASSICAL GEOGRAPHY. By H. F. TOZER, M.A.
  GEOGRAPHY. By Sir G. GROVE, D.C.L. Maps.
  ROMAN ANTIQUITIES. By Prof. WILKINS, Litt.D. Illustrated.

ANALYSIS OF ENGLISH HISTORY. By Prof. T. F. TOUT, M.A.
INDIAN HISTORY: ASIATIC AND EUROPEAN. By J. TALBOYS WHEELER.
HOLE.—A GENEALOGICAL STEMMA OF THE KINGS OF ENGLAND AND FRANCE. By Rev. C. HOLE. On Sheet. 1s.
JENNINGS.—CHRONOLOGICAL TABLES. A synchronistic arrangement of the events of Ancient History (with an Index). By Rev. ARTHUR C. JENNINGS. 8vo. 5s.
LABBERTON.—NEW HISTORICAL ATLAS AND GENERAL HISTORY. By R. H. LABBERTON. 4to. New Ed., revised and enlarged. 15s.
LETHBRIDGE.—A SHORT MANUAL OF THE HISTORY OF INDIA. With an Account of INDIA AS IT IS. The Soil, Climate, and Productions; the People, their Races, Religions, Public Works, and Industries; the Civil Services, and System of Administration. By Sir ROPER LETHBRIDGE, Fellow of the Calcutta University. With Maps. Cr. 8vo. 5s.

D

MAHAFFY.—GREEK LIFE AND THOUGHT FROM THE AGE OF ALEXANDER TO THE ROMAN CONQUEST. By Rev. J. P. Mahaffy, D.D., Fellow of Trinity College, Dublin. Cr. 8vo. 12s. 6d.
   THE GREEK WORLD UNDER ROMAN SWAY. From Plutarch to Polybius. By the same Author. Cr. 8vo. 10s. 6d.
MARRIOTT.—THE MAKERS OF MODERN ITALY: Mazzini, Cavour, Garibaldi. Three Lectures. By J. A. R. Marriott, M.A., Lecturer in Modern History and Political Economy, Oxford. Cr. 8vo. 1s. 6d.
MATHEW.—HISTORY READERS FOR ELEMENTARY SCHOOLS. Adapted to the several Standards. Edited by Edward J. Mathew.
MICHELET.—A SUMMARY OF MODERN HISTORY. By M. Michelet. Translated by M. C. M. Simpson. Gl. 8vo. 4s. 6d.
NORGATE.—ENGLAND UNDER THE ANGEVIN KINGS. By Kate Norgate. With Maps and Plans. 2 vols. 8vo. 32s.
OTTÉ.—SCANDINAVIAN HISTORY. By E. C. Otté. With Maps. Gl. 8vo. 6s.
SEELEY.—Works by J. R. Seeley, M.A., Regius Professor of Modern History in the University of Cambridge.
   THE EXPANSION OF ENGLAND. Crown 8vo. 4s. 6d.
   OUR COLONIAL EXPANSION. Extracts from the above. Cr. 8vo. Sewed. 1s.
*****TAIT.**—ANALYSIS OF ENGLISH HISTORY, based on Green's "Short History of the English People." By C. W. A. Tait, M.A., Assistant Master at Clifton. Revised and Enlarged Ed. Cr. 8vo. 4s. 6d.
WHEELER.—Works by J. Talboys Wheeler.
   *A PRIMER OF INDIAN HISTORY. Asiatic and European. 18mo. 1s.
   *COLLEGE HISTORY OF INDIA, ASIATIC AND EUROPEAN. With Maps. Cr. 8vo. 3s.; sewed, 2s. 6d.
   A SHORT HISTORY OF INDIA AND OF THE FRONTIER STATES OF AFGHANISTAN, NEPAUL, AND BURMA. With Maps. Cr. 8vo. 12s.
YONGE.—Works by Charlotte M. Yonge.
   CAMEOS FROM ENGLISH HISTORY. Ex. fcap. 8vo. 5s. each. (1) FROM ROLLO TO EDWARD II. (2) THE WARS IN FRANCE. (3) THE WARS OF THE ROSES. (4) REFORMATION TIMES. (5) ENGLAND AND SPAIN. (6) FORTY YEARS OF STUART RULE (1603-1643). (7) REBELLION AND RESTORATION (1642-1678).
   EUROPEAN HISTORY. Narrated in a Series of Historical Selections from the Best Authorities. Edited and arranged by E. M. Sewell and C. M. Yonge. Cr. 8vo. First Series, 1003-1154. 6s. Second Series, 1088-1228. 6s.
   THE VICTORIAN HALF CENTURY—A JUBILEE BOOK. With a New Portrait of the Queen. Cr. 8vo. Paper covers, 1s. Cloth, 1s. 6d.

ART.

*****ANDERSON.**—LINEAR PERSPECTIVE AND MODEL DRAWING. A School and Art Class Manual, with Questions and Exercises for Examination, and Examples of Examination Papers. By Laurence Anderson. Illustrated. 8vo. 2s.
COLLIER.—A PRIMER OF ART. By the Hon. John Collier. Illustrated. 18mo. 1s.
COOK.—THE NATIONAL GALLERY, A POPULAR HANDBOOK TO. By Edward T. Cook, with a preface by John Ruskin, LL.D., and Selections from his Writings. 3d Ed. Cr. 8vo. Half-morocco, 14s.
   ⁎ Also an Edition on large paper, limited to 250 copies. 2 vols. 8vo.
DELAMOTTE.—A BEGINNER'S DRAWING BOOK. By P. H. Delamotte, F.S.A. Progressively arranged. New Ed., improved. Cr. 8vo. 3s. 6d.
ELLIS.—SKETCHING FROM NATURE. A Handbook for Students and Amateurs. By Tristram J. Ellis. Illustrated by H. Stacy Marks, R.A., and the Author. New Ed., revised and enlarged. Cr. 8vo. 3s. 6d.

GROVE.—A DICTIONARY OF MUSIC AND MUSICIANS. A.D. 1450-1889. Edited by Sir GEORGE GROVE, D.C.L. In four vols. 8vo. Price 21s. each. Also in Parts.
Parts I.-XIV., Parts XIX.-XXII., 8s. 6d. each. Parts XV., XVI., 7s. Parts XVII., XVIII., 7s. Parts XXIII.-XXV. (Appendix), 9s.
A COMPLETE INDEX TO THE ABOVE. By Mrs. E. WODEHOUSE. 8vo. 7s. 6d.
HUNT.—TALKS ABOUT ART. By WILLIAM HUNT. With a Letter from Sir J. E. MILLAIS, Bart., R.A. Cr. 8vo. 3s. 6d.
MELDOLA.—THE CHEMISTRY OF PHOTOGRAPHY. By RAPHAEL MELDOLA, F.R.S., Professor of Chemistry in the Technical College, Finsbury. Cr. 8vo. 6s.
TAYLOR.—A PRIMER OF PIANOFORTE-PLAYING. By FRANKLIN TAYLOR. Edited by Sir GEORGE GROVE. 18mo. 1s.
TAYLOR.—A SYSTEM OF SIGHT-SINGING FROM THE ESTABLISHED MUSICAL NOTATION; based on the Principle of Tonic Relation, and Illustrated by Extracts from the Works of the Great Masters. By SEDLEY TAYLOR. 8vo. 5s. net.
TYRWHITT.—OUR SKETCHING CLUB. Letters and Studies on Landscape Art. By Rev. R. ST. JOHN TYRWHITT. With an authorised Reproduction of the Lessons and Woodcuts in Prof. Ruskin's "Elements of Drawing.". 5th Ed. Cr. 8vo. 7s. 6d.

DIVINITY.

ABBOTT.—BIBLE LESSONS. By Rev. EDWIN A. ABBOTT, D.D. Cr. 8vo. 4s. 6d.
ABBOTT—RUSHBROOKE.—THE COMMON TRADITION OF THE SYNOPTIC GOSPELS, in the Text of the Revised Version. By Rev. EDWIN A. ABBOTT, D.D., and W. G. RUSHBROOKE, M.L. Cr. 8vo. 3s. 6d.
ARNOLD.—Works by MATTHEW ARNOLD.
A BIBLE-READING FOR SCHOOLS,—THE GREAT PROPHECY OF ISRAEL'S RESTORATION (Isaiah, Chapters xl.-lxvi.) Arranged and Edited for Young Learners. 18mo. 1s.
ISAIAH XL-LXVI. With the Shorter Prophecies allied to it. Arranged and Edited, with Notes. Cr. 8vo. 5s.
ISAIAH OF JERUSALEM, IN THE AUTHORISED ENGLISH VERSION. With Introduction, Corrections and Notes. Cr. 8vo. 4s. 6d.
BENHAM.—A COMPANION TO THE LECTIONARY. Being a Commentary on the Proper Lessons for Sundays and Holy Days. By Rev. W. BENHAM, B.D. Cr. 8vo. 4s. 6d.
CASSEL.—MANUAL OF JEWISH HISTORY AND LITERATURE; preceded by a BRIEF SUMMARY OF BIBLE HISTORY. By Dr. D. CASSEL. Translated by Mrs. H. LUCAS. Fcap. 8vo. 2s. 6d.
CHURCH.—STORIES FROM THE BIBLE. By Rev. A. J. CHURCH, M.A. Illustrated. 2 parts. Cr. 8vo. 3s. 6d. each.
*CROSS.—BIBLE READINGS SELECTED FROM THE PENTATEUCH AND THE BOOK OF JOSHUA. By Rev. JOHN A. CROSS. 2d Ed., enlarged, with Notes. Gl. 8vo. 2s. 6d.
DRUMMOND.—INTRODUCTION TO THE STUDY OF THEOLOGY. By JAMES DRUMMOND, LL.D., Professor of Theology in Manchester New College, London. Cr. 8vo. 5s.
FARRAR.—Works by the Venerable Archdeacon F. W. FARRAR, D.D., F.R.S., Archdeacon and Canon of Westminster.
THE HISTORY OF INTERPRETATION. Bampton Lectures, 1885. 8vo. 16s.
THE MESSAGES OF THE BOOKS. Being Discourses and Notes on the Books of the New Testament. 8vo. 14s.
*GASKOIN.—THE CHILDREN'S TREASURY OF BIBLE STORIES. By Mrs. HERMAN GASKOIN. Edited with Preface by Rev. G. F. MACLEAR, D.D. 18mo. 1s. each. Part I.—OLD TESTAMENT HISTORY. Part II.—NEW TESTAMENT. Part III.—THE APOSTLES: ST. JAMES THE GREAT, ST. PAUL, AND ST. JOHN THE DIVINE.

GOLDEN TREASURY PSALTER.—Students' Edition. Being an Edition of "The Psalms chronologically arranged, by Four Friends," with briefer Notes. 18mo. 3s. 6d.

GREEK TESTAMENT.—Edited, with Introduction and Appendices, by Bishop WESTCOTT and Dr. F. J. A. HORT. Two Vols. Cr. 8vo. 10s. 6d. each. Vol. I. The Text. Vol. II. Introduction and Appendix.
SCHOOL EDITION OF TEXT. 12mo. Cloth, 4s. 6d.; Roan, red edges, 5s. 6d. 18mo. Morocco, gilt edges, 6s. 6d.
*GREEK TESTAMENT, SCHOOL READINGS IN THE. Being the outline of the life of our Lord, as given by St. Mark, with additions from the Text of the other Evangelists. Arranged and Edited, with Notes and Vocabulary, by Rev. A. CALVERT, M.A. Fcap. 8vo. 2s. 6d.
*THE GOSPEL ACCORDING TO ST. MATTHEW. Being the Greek Text as revised by Bishop WESTCOTT and Dr. HORT. With Introduction and Notes by Rev. A. SLOMAN, M.A., Headmaster of Birkenhead School. Fcap. 8vo. 2s. 6d.
THE GOSPEL ACCORDING TO ST. MARK. Being the Greek Text as revised by Bishop WESTCOTT and Dr. HORT. With Introduction and Notes by Rev. J. O. F. MURRAY, M.A., Lecturer at Emmanuel College, Cambridge. Fcap. 8vo. [*In preparation.*]
*THE GOSPEL ACCORDING TO ST. LUKE. Being the Greek Text as revised by Bishop WESTCOTT and Dr. HORT. With Introduction and Notes by Rev. JOHN BOND, M.A. Fcap. 8vo. 2s. 6d.
*THE ACTS OF THE APOSTLES. Being the Greek Text as revised by Bishop WESTCOTT and Dr. HORT. With Explanatory Notes by T. E. PAGE, M.A., Assistant Master at the Charterhouse. Fcap. 8vo. 3s. 6d.

GWATKIN.—CHURCH HISTORY TO THE BEGINNING OF THE MIDDLE AGES. By H. M. GWATKIN, M.A. 8vo. [*In preparation.*]

HARDWICK.—Works by Archdeacon HARDWICK.
A HISTORY OF THE CHRISTIAN CHURCH. Middle Age. From Gregory the Great to the Excommunication of Luther. Edited by W. STUBBS, D.D., Bishop of Oxford. With 4 Maps. Cr. 8vo. 10s. 6d.
A HISTORY OF THE CHRISTIAN CHURCH DURING THE REFORMATION. 9th Ed. Edited by Bishop STUBBS. Cr. 8vo. 10s. 6d.

HOOLE.—THE CLASSICAL ELEMENT IN THE NEW TESTAMENT. Considered as a proof of its Genuineness, with an Appendix on the Oldest Authorities used in the Formation of the Canon. By CHARLES H. HOOLE, M.A., Student of Christ Church, Oxford. 8vo. 10s. 6d.

JENNINGS and LOWE.—THE PSALMS, WITH INTRODUCTIONS AND CRITICAL NOTES. By A. C. JENNINGS, M.A.; assisted in parts by W. H. LOWE, M.A. In 2 vols. 2d Ed., revised. Cr. 8vo. 10s. 6d. each.

KIRKPATRICK.—THE MINOR PROPHETS. Warburtonian Lectures. By Rev. Prof. KIRKPATRICK. [*In preparation.*]
THE DIVINE LIBRARY OF THE OLD TESTAMENT. By the same. [*In prep.*]

KUENEN.—PENTATEUCH AND BOOK OF JOSHUA: An Historico-Critical Inquiry into the Origin and Composition of the Hexateuch. By A KUENEN. Translated by P. H. WICKSTEED, M.A. 8vo. 14s.

LIGHTFOOT.—Works by the Right Rev. J. B. LIGHTFOOT, D.D., late Bishop of Durham.
ST. PAUL'S EPISTLE TO THE GALATIANS. A Revised Text, with Introduction, Notes, and Dissertations. 10th Ed., revised. 8vo. 12s.
ST. PAUL'S EPISTLE TO THE PHILIPPIANS. A Revised Text, with Introduction, Notes, and Dissertations. 9th Ed., revised. 8vo. 12s.
ST. PAUL'S EPISTLES TO THE COLOSSIANS AND TO PHILEMON. A Revised Text, with Introductions, Notes, and Dissertations. 8th Ed., revised. 8vo. 12s.
THE APOSTOLIC FATHERS. Part I. ST. CLEMENT OF ROME. A Revised Text, with Introductions, Notes, Dissertations, and Translations. 2 vols. 8vo. 32s.
THE APOSTOLIC FATHERS. Part II. ST. IGNATIUS—ST. POLYCARP. Revised Texts, with Introductions, Notes, Dissertations, and Translations. 2d Ed. 3 vols. 8vo. 48s.

DIVINITY

THE APOSTOLIC FATHERS. Abridged Edition. With short Introductions, Greek Text, and English Translation. 8vo. 16s.

ESSAYS ON THE WORK ENTITLED "SUPERNATURAL RELIGION." (Reprinted from the *Contemporary Review*.) 8vo. 10s. 6d.

MACLEAR.—Works by the Rev. G. F. MACLEAR, D.D., Warden of St. Augustine's College, Canterbury.

ELEMENTARY THEOLOGICAL CLASS-BOOKS.

*A SHILLING BOOK OF OLD TESTAMENT HISTORY. With Map. 18mo.

*A SHILLING BOOK OF NEW TESTAMENT HISTORY. With Map. 18mo. These works have been carefully abridged from the Author's large manuals.

*A CLASS-BOOK OF OLD TESTAMENT HISTORY. Maps. 18mo. 4s. 6d.

*A CLASS-BOOK OF NEW TESTAMENT HISTORY, including the Connection of the Old and New Testaments. With maps. 18mo. 5s. 6d.

AN INTRODUCTION TO THE THIRTY-NINE ARTICLES. [*In the Press.*

*AN INTRODUCTION TO THE CREEDS. 18mo. 2s. 6d.

*A CLASS-BOOK OF THE CATECHISM OF THE CHURCH OF ENGLAND. 18mo. 1s. 6d.

*A FIRST CLASS-BOOK OF THE CATECHISM OF THE CHURCH OF ENGLAND. With Scripture Proofs. 18mo. 6d.

*A MANUAL OF INSTRUCTION FOR CONFIRMATION AND FIRST COMMUNION. WITH PRAYERS AND DEVOTIONS. 32mo. 2s.

MAURICE.—THE LORD'S PRAYER, THE CREED, AND THE COMMANDMENTS. To which is added the Order of the Scriptures. By Rev. F. D. MAURICE, M.A. 18mo. 1s.

THE PENTATEUCH AND BOOK OF JOSHUA: An Historico-Critical Inquiry into the Origin and Composition of the Hexateuch. By A. KUENEN, Professor of Theology at Leiden. Translated by P. H. WICKSTEED, M.A. 8vo. 14s.

PROCTER.—A HISTORY OF THE BOOK OF COMMON PRAYER, with a Rationale of its Offices. By Rev. F. PROCTER. 18th Ed. Cr. 8vo. 10s. 6d.

*PROCTER and MACLEAR.—AN ELEMENTARY INTRODUCTION TO THE BOOK OF COMMON PRAYER. Rearranged and supplemented by an Explanation of the Morning and Evening Prayer and the Litany. By Rev. F. PROCTER and Rev. Dr. MACLEAR. New Edition, containing the Communion Service and the Confirmation and Baptismal Offices. 18mo. 2s. 6d.

THE PSALMS, CHRONOLOGICALLY ARRANGED. By Four Friends. New Ed. Cr. 8vo. 5s. net.

THE PSALMS, WITH INTRODUCTIONS AND CRITICAL NOTES. By A. C. JENNINGS, M.A., Jesus College, Cambridge; assisted in parts by W. H. LOWE, M.A., Hebrew Lecturer at Christ's College, Cambridge. In 2 vols. 2d Ed., revised. Cr. 8vo. 10s. 6d. each.

RYLE.—AN INTRODUCTION TO THE CANON OF THE OLD TESTAMENT. By Rev. H. E. RYLE, M.A., Hulsean Professor of Divinity in the University of Cambridge. Cr. 8vo. [*In preparation.*

SIMPSON.—AN EPITOME OF THE HISTORY OF THE CHRISTIAN CHURCH DURING THE FIRST THREE CENTURIES, AND OF THE REFORMATION IN ENGLAND. By Rev. WILLIAM SIMPSON, M.A. 7th Ed. Fcap. 8vo. 3s. 6d.

ST. JAMES' EPISTLE.—The Greek Text, with Introduction and Notes. By Rev. JOSEPH MAYOR, M.A., Professor of Moral Philosophy in King's College, London. 8vo. [*In the Press.*

ST. JOHN'S EPISTLES.—The Greek Text, with Notes and Essays. By Right Rev. B. F. WESTCOTT, D.D., Bishop of Durham. 2d Ed., revised. 8vo. 12s. 6d.

ST. PAUL'S EPISTLES.—THE EPISTLE TO THE ROMANS. Edited by the Very Rev. C. J. VAUGHAN, D.D., Dean of Llandaff. 5th Ed. Cr. 8vo. 7s. 6d.

THE TWO EPISTLES TO THE CORINTHIANS, A COMMENTARY ON. By the late Rev. W. KAY, D.D., Rector of Great Leghs, Essex. 8vo. 9s.

THE EPISTLE TO THE GALATIANS. Edited by the Right Rev. J. B. LIGHTFOOT, D.D. 10th Ed. 8vo. 12s.

THE EPISTLE TO THE PHILIPPIANS. By the Same Editor. 9th Ed. 8vo. 12s.

THE EPISTLE TO THE PHILIPPIANS, with Translation, Paraphrase, and Notes for English Readers. By the Very Rev. C. J. VAUGHAN, D.D. Cr. 8vo. 5s.

THE EPISTLE TO THE COLOSSIANS AND TO PHILEMON. By the Right Rev. J. B. LIGHTFOOT, D.D. 8th Ed. 8vo. 12s.

THE EPISTLES TO THE EPHESIANS, THE COLOSSIANS, AND PHILEMON; with Introductions and Notes, and an Essay on the Traces of Foreign Elements in the Theology of these Epistles. By Rev. J. LLEWELYN DAVIES, M.A. 8vo. 7s. 6d.

THE EPISTLE TO THE THESSALONIANS, COMMENTARY ON THE GREEK TEXT. By JOHN EADIE, D.D. Edited by Rev. W. YOUNG, M.A., with Preface by Prof. CAIRNS. 8vo. 12s.

THE EPISTLE TO THE HEBREWS.—In Greek and English. With Critical and Explanatory Notes. Edited by Rev. F. RENDALL, M.A. Cr. 8vo. 6s.

THE ENGLISH TEXT, WITH COMMENTARY. By the same Editor. Cr. 8vo. 7s. 6d.

THE GREEK TEXT. With Notes by C. J. VAUGHAN, D.D., Dean of Llandaff. Cr. 8vo. 7s. 6d.

THE GREEK TEXT. With Notes and Essays by the Right Rev. Bishop WESTCOTT, D.D. 8vo. 14s.

VAUGHAN.—THE CHURCH OF THE FIRST DAYS. Comprising the Church of Jerusalem, the Church of the Gentiles, the Church of the World. By C. J. VAUGHAN, D.D., Dean of Llandaff. New Ed. Cr. 8vo. 10s. 6d.

WESTCOTT.—Works by the Right Rev. BROOKE FOSS WESTCOTT, D.D., Bishop of Durham.

A GENERAL SURVEY OF THE HISTORY OF THE CANON OF THE NEW TESTAMENT DURING THE FIRST FOUR CENTURIES. 6th Ed. With Preface on "Supernatural Religion." Cr. 8vo. 10s. 6d.

INTRODUCTION TO THE STUDY OF THE FOUR GOSPELS. 7th Ed. Cr. 8vo. 10s. 6d.

THE BIBLE IN THE CHURCH. A Popular Account of the Collection and Reception of the Holy Scriptures in the Christian Churches. 18mo. 4s. 6d.

THE EPISTLES OF ST. JOHN. The Greek Text, with Notes and Essays. 2d Ed., revised. 8vo. 12s. 6d.

THE EPISTLE TO THE HEBREWS. The Greek Text, with Notes and Essays. 8vo. 14s.

SOME THOUGHTS FROM THE ORDINAL. Cr. 8vo. 1s. 6d.

WESTCOTT and HORT.—THE NEW TESTAMENT IN THE ORIGINAL GREEK. The Text, revised by the Right Rev. Bishop WESTCOTT and Dr. F. J. A. HORT. 2 vols. Cr. 8vo. 10s. 6d. each. Vol. I. Text. Vol. II. Introduction and Appendix.

SCHOOL EDITION OF TEXT. 12mo. 4s. 6d.; Roan, red edges, 5s. 6d. Fcap. 8vo. Morocco, gilt edges, 6s. 6d.

WRIGHT.—THE COMPOSITION OF THE FOUR GOSPELS. A Critical Enquiry. By Rev. ARTHUR WRIGHT, M.A., Fellow and Tutor of Queen's College, Cambridge. Cr. 8vo. 5s.

WRIGHT.—THE BIBLE WORD-BOOK: A Glossary of Archaic Words and Phrases in the Authorised Version of the Bible and the Book of Common Prayer. By W. ALDIS WRIGHT, M.A., Vice-Master of Trinity College, Cambridge. 2d Ed., revised and enlarged. Cr. 8vo. 7s. 6d.

*****YONGE.**—SCRIPTURE READINGS FOR SCHOOLS AND FAMILIES. By CHARLOTTE M. YONGE. In Five Vols. Ex. fcap. 8vo. 1s. 6d. each. With Comments. 3s. 6d. each.

FIRST SERIES.—GENESIS TO DEUTERONOMY. SECOND SERIES.—FROM JOSHUA TO SOLOMON THIRD SERIES.—THE KINGS AND THE PROPHETS. FOURTH SERIES. —THE GOSPEL TIMES. FIFTH SERIES.—APOSTOLIC TIMES.

ZECHARIAH—THE HEBREW STUDENT'S COMMENTARY ON ZECHARIAH, HEBREW AND LXX. With Excursus on Syllable-dividing, Metheg, Initial Dagesh, and Siman Rapheh. By W. H. LOWE, M.A., Hebrew Lecturer at Christ's College, Cambridge. 8vo. 10s. 6d.

The English Illustrated Magazine.

Each Volume Complete in Itself.

Volume for 1884.

Containing 792 pages, with 428 Illustrations. Price 7s. 6d.

The Volume contains the following Complete Stories and Serials:—

The Armourer's 'Prentices. By C. M. YONGE. **An Unsentimental Journey through Cornwall.** By Mrs. CRAIK. **Julia.** By WALTER BESANT. **How I became a War Correspondent.** By ARCHIBALD FORBES. **The Story of a Courtship.** By STANLEY J. WEYMAN, etc.

Volume for 1885.

Containing 840 pages, with nearly 500 Illustrations. Price 8s.

The Volume contains the following Complete Stories and Serials:—

A Family Affair. By HUGH CONWAY. **Girl at the Gate.** By WILKIE COLLINS. **The Path of Duty.** By HENRY JAMES. **Schwartz.** By D. CHRISTIE MURRAY. **A Ship of '49.** By BRET HARTE. **That Terrible Man.** By W. E. NORRIS. **Interviewed by an Emperor.** By ARCHIBALD FORBES. **In the Lion's Den.** By the Author of "John Herring," etc.

Volume for 1886.

Containing 832 pages, with nearly 500 Illustrations. Price 8s.

The Volume contains the following Complete Stories and Serials:—

Kiss and be Friends. By the Author of "John Halifax, Gentleman." **Aunt Rachel.** By D. CHRISTIE MURRAY. **A Garden of Memories.** By MARGARET VELEY. **My Friend Jim.** By W. E. NORRIS. **Harry's Inheritance.** By GRANT ALLEN. **Captain Lackland.** By CLEMENTINA BLACK. **Witnessed by Two.** By Mrs. MOLESWORTH. **The Poetry did It.** By WILKIE COLLINS. **Dr. Barrere.** By Mrs. OLIPHANT. **Mere Suzanne.** By KATHARINE S. MACQUOID. **Days with Sir Roger de Coverley,** with pictures by HUGH THOMSON, etc.

Volume for 1887.

Containing 832 pages, with nearly 500 Illustrations. Price 8s.

The Volume contains the following Complete Stories and Serials:—

Marzio's Crucifix. By F. MARION CRAWFORD. **A Secret Inheritance.** By B. L. FARJEON. **Jacquetta.** By the Author of "John Herring." **Gerald.** By STANLEY J. WEYMAN. **An Unknown Country.** By the Author of "John Halifax, Gentleman." With Illustrations by F. NOEL PATON. **A Siege Baby.** By J. S. WINTER. **Miss Falkland.** By CLEMENTINA BLACK, etc.

The English Illustrated Magazine—*continued*.

Volume for 1888.

Containing 832 pages, with nearly 500 Illustrations. Price 8s.

Among the chief Contents of the Volume are the following Complete Stories and Serials :—

Coaching Days and Coaching Ways. By W. O. TRISTRAM. With Illustrations by H. RAILTON and HUGH THOMSON. **The Story of Jael.** By the Author of "Mehalah." **Lil: a Liverpool Child.** By AGNES C. MAITLAND. **The Patagonia.** By HENRY JAMES. **Family Portraits.** By S. J. WEYMAN. **The Mediation of Ralph Hardelot.** By Prof. W. MINTO. **That Girl in Black.** By Mrs. MOLESWORTH. **Glimpses of Old English Homes.** By ELIZABETH BALCH. **Pagodas, Aurioles, and Umbrellas.** By C. F. GORDON CUMMING. **The Magic Fan.** By JOHN STRANGE WINTER.

Volume for 1889.

Containing 900 pages, with nearly 500 Illustrations. Price 8s.

Among the chief Contents of the Volume are the following Complete Stories and Serials :—

Sant' Ilario. By F. MARION CRAWFORD. **The House of the Wolf.** By STANLEY J. WEYMAN. **Glimpses of Old English Homes.** By ELIZABETH BALCH. **One Night—The Better Man.** By ARTHUR PATERSON. **How the "Crayture" got on the Strength.** And other Sketches. By ARCHIBALD FORBES. **La Belle Americaine.** By W. E. NORRIS. **Success.** By KATHARINE S. MACQUOID. **Jenny Harlowe.** By W. CLARK RUSSELL.

Volume for 1890.

Containing 900 pages, with nearly 550 Illustrations. Price 8s.

Among the chief Contents of the Volume are the following Complete Stories and Serials :—

The Ring of Amasis. By the EARL OF LYTTON. **The Glittering Plain: or, the Land of Living Men.** By WILLIAM MORRIS. **The Old Brown Mare.** By W. E. NORRIS. **My Journey to Texas.** By ARTHUR PATERSON. **A Glimpse of Highclere Castle—A Glimpse of Osterley Park.** By ELIZABETH BALCH. **For the Cause.** By STANLEY J. WEYMAN. **Morised.** By the MARCHIONESS OF CARMARTHEN. **Overland from India.** By Sir DONALD MACKENZIE WALLACE, K.C.I.E. **The Doll's House and After.** By WALTER BESANT. **La Mulette, Anno 1814.** By W. CLARK RUSSELL.

Volume for 1891.

Containing 900 pages, and about 500 Illustrations. Price 8s.

Among the chief Contents of the Volume are the following Complete Stories and Serials :—

The Witch of Prague. By F. MARION CRAWFORD. **The Wisdom Tooth.** By D. CHRISTIE MURRAY and HENRY HERMAN. **Wooden Tony.** By Mrs. W. K. CLIFFORD. **Two Jealousies.** By ALAN ADAIR. **Gentleman Jim.** By MARY GAUNT. **Harrow School. Winchester College. Fawsley Park. Ham House. Westminster Abbey. Norwich. The New Trade-Union Movement. Russo-Jewish Immigrant. Queen's Private Garden at Osborne.**

MACMILLAN AND CO., LONDON.

VII.50.10.91.

www.ingramcontent.com/pod-product-compliance
Lightning Source LLC
Chambersburg PA
CBHW032057220426
43664CB00008B/1039